Simply Learning, Simply Best

Simply Learning, Simply Best

國貿一魚兩吃 新多益

考用」與「職場」大結合
速掌握「商用英文書信」要訣！

婉婷◎著

30篇精選國貿英文信件與15回必考新多益閱讀

考試高分與職場升遷一次GET

NEW TOEIC閱讀高分，雙篇閱讀是高分分水嶺

令老闆、客戶驚艷，撰寫適切商用英文書信是關鍵

一本於撰寫職場國貿英文信件與新多益閱讀的**強效結合**

有效提升職場英文信件溝通，E-mail英文書信往來**不凸槌**

立即提升新多益閱讀高分，新多益奪金**好容易!!!**

作者序

記得自己是在旅居美國九年後，在對 NEW TOEIC 的好奇及“金色證書”的誘惑下報考了 NEW TOEIC。看到考題後我驚訝了！這樣生活化的英語，如果不是像我們多年融入美國人的生活，一般台灣考生怎麼可能瞭解。NEW TOEIC 考的可不只是英文，而是在生活中運用英語的流利度。果不其然，轉頭看看隔壁的學生，個個搔頭皺眉。回想起自己剛開始在美國工作時，也是為了寫出專業卻不生硬的商用英語文書，包含書信，契約，廣告等，搞得焦頭爛額。

正因這樣的理由，激起了將 NEW TOEIC 中的雙篇閱讀及商用英語文書做一個結合。英文表達要道地，就一定要瞭解當地人士平常是怎麼使用的。在這本書裡，沒有咬文嚼字，沒有賣弄文法，你可以輕鬆的學習到英美人士日常生活的語言及書信的寫作，簡單並清楚的瞭解文章的內容並輕鬆的寫出自己想表達的文書。相信對於 NEW TOEIC 閱讀的拿分及商用書信的寫作都會有相當的幫助。

洪婉婷

目　次

國貿主題篇 International Trade

新多益雙篇閱讀篇 NEW TOEIC Double Passages

International Trade

Unit 1 邀請 Invitation

1-1 新商品代理商發表會 New Product Release Announcement to Representatives

All **respectful representatives**,

It gives us great pleasure to invite you to our headquarter to join the new product line release party.

As we know the **demand** for coffee is increasing worldwide, being one of the biggest vendos in the industry, BestCoffee has spent a great amount of time researching a better way to create decaf coffee beans which replace the chemical wash with other natural process to **eliminate** possible chemical damage to human body. It yet **maintains** the rich coffee aroma and gives our decaf customers a much better **satisfaction**.

In this product line, we will not only **release** the standard decaf coffee beans, but also the flavored decaf coffee beans, such as caramel, coconuts, mocha, and such. The variety of flavors will give the coffee shops a full menu of flavored coffee with decaf options.

With all the excited news, BestCoffee would like to invite all our representatives to join the product pre-release party on Nov 1st, 2014. During the party, we will not only be offering you the tasting for all BestCoffee products, but also go thru the **manufacturing** process for the new product line. This will give all of you a good idea about the difference between the decaf coffee bean made by BestCoffee and other companies. The actual release date for this product line is currently set on Dec. 25th. This should give all our representatives enough time to get the demand for the holiday sale after Christmas.

Please respond back to us in this regard at your earliest convenience. We will make the hotel reservations and traffic arrangements accordingly.

Yours sincerely,
Debra McDonald
General Manager
BestCoffee, Inc.

中譯

所有恭敬的代表，

我們非常榮幸地邀請您參加我們總部即將舉辦的新產品線發布派對。

由於我們瞭解全球的咖啡需求不斷增加，且為該行業最大的供應商之一，BestCoffee 花費大量的時間研究開發一種更好的方式來製造低咖啡因咖啡豆，以自然的方式取代化學清洗，以消除人體受到可能的化學傷害。不但如此，它還能保持著濃郁的咖啡香氣，使我們的低咖啡因客戶也能得到一個更好的滿意度。

在這條產品線裡，我們不但會發表標準低咖啡因的咖啡豆，同時也會發表低咖啡因的風味咖啡豆，如焦糖、椰子、摩卡等。各種口味的提供將會讓各咖啡館有一個完整的低咖啡因菜單選項。

由於這興奮的消息，BestCoffee 希望能邀請我們所有的代理商參加在 2014 年 11 月 1 日該產品的預定發表會，我們不僅將提供您品嚐所有 BestCoffee 的產品，也會帶領您瞭解新生產線的製造過程。這個過程將會讓您們了解本公司的低咖啡因咖啡豆和其他公司的咖啡因咖啡豆生產之間的差異。該生產線的實際發布日期是在 12 月 25 日。這次的發表會應該給與我們所有的代理商足夠的時間準備聖誕節前後銷售的需求。

請在您方便的狀況下儘快回覆給我們是否參加。我們將會依不同需求預訂酒店和交通安排。

此致，

Debra McDonald

總經理

BestCoffee 公司

達人提點 I

在這篇文章中，我們集合了兩個要點。(1) 新產品發表 (2) 邀請信函。

(1) 新產品發表的部份，想當然我們在文章的開頭要先說明背景 background，即是為什麼“Why”需要這個產品。再者，我們需要說明這個新產品的特點，也就是“What is special ? ”。說明這項新產品與市面上既有的產品的不同，這樣才能提起閱讀者的興趣。最後則是列出這項新產品所有的品項，以告訴讀者 what are available。

(2) 邀請信函的部份也是一樣，在本文的開頭應先說明邀請的原因“Why”。範例的“新產品發表”部份即為說明為什麼寄送邀請信函的目的。接下來就應該明確說明宴會的時間及地點，並告知內容及流程。最後，當然是希望對方能出席，並給與出席與否的回函。

字彙

1. **Respectful**: (adj) 恭敬的；尊敬人的，尊重人的

   When entering the historical site, such as Peal Harbor Memorial, it is important to keep a respectful silence.

   當訪問如珍珠港的古蹟時，保持沉默以表示恭敬是很重要的。

2. **Representative**: (n) 代表，代理人，代理商

   He is attending the convention as the company's representative.

   他是以公司代理商的身分參加會議。

3. **Demand**: (n) 需要，需求

   Due to the bad weather condition, the supply of fruit falls short of demand this year.

   由於壞天氣的關係 , 今年水果供不應求。

4. **Eliminate**: (v) 排除，消除，消滅

   One of the government's big goals for 10 years is to eliminate poverty.

   政府十年大目標中，其中一項是消滅貧困。

5. **Maintain**: (v) 維持；保持；使繼續

   Peter and David have been maintaining their friendship even after marriage.

   Peter 和 David 在婚後也一直保持著他們之間的友誼。

6. **Satisfaction**: (n) 滿意，滿足；稱心

He smiled in satisfaction when he finished the science project.

他完成科學計畫後滿意地笑了。

7. **Release**: (v) 發行；發表

Fans are excited that a new cell phone is to be released today.

粉絲們都很興奮今天要發行一款新的手機。

8. **Manufacture**: (v) 製造，加工

The well-known German company decided to move their certain car models to be manufactured in Mexico.

某知名德國公司將旗下的一些車款移至墨西哥製造。

1-2 邀請函回覆並要求其他會議 Invitation Response and Meeting Request

Dear Ms. McDonald,

It will be our pleasure to attend the product pre-release party. As you mentioned in the invitation, the demand for coffee is increasing **globally**. Especially in Asia **region**, we are seeing a dramatic increase in demand for flavored and decaf coffee beans because it has become a **trend** to enjoy a creative coffee in the early morning everyday for all ages. As much as we are concerned about teens enjoying coffee which might **affect** their growth, I believe the all-natural decaf coffee been will be a great hit for all age groups.

While attending the pre-release party, I am hoping we can **arrange** a couple of private meetings with your **supply** group and the QA group. The reason for that is because our government has recently changed the import law, which demands a longer **lead-time** for shipments and a longer expiration time for all food products imported to our country. There are several concerns we have due to this matter and would like to discuss with your groups during our visit.

Please kindly let us know if your groups will be available and we will arrange our visit time **accordingly**. In regards to the hotel and traffic arrangements, I very much appreciate your offer, but I will have my assistant Debby take care of them from here since we might be visiting several other companies while we are in the area.

We look forward to meeting with you in November.

Sincerely yours,
Aden Cheng
Sales Head,
Republic of Coffee Nation

中譯

尊敬的 McDonald 女士，

這將是我們的榮幸來參加產品發表派對。正如您在邀請中提到，全球的咖啡需求不斷增加，特別是在亞洲地區，我們看到的風味咖啡和低咖啡因咖啡豆的需求急劇增加，這是在於各年齡層的消費者每天早晨享受一杯創意咖啡已成為一種趨勢。在同時，正如我們關心青少年享受咖啡時可能會對成長造成的影響，我相信全天然無咖啡因的咖啡將受到所有年齡層的歡迎。

在出席發表會的同時，我希望我們可以與您的供應組及品管組安排幾個私人會議。原因是因為我們的政府最近改變了進口法，這個改變使得進口至我國食品需要一個較長的交貨時間，也要求所有食品需要有較長的保存期限。由於這個改變，有幾個我們關注的問題希望可以藉由這次的訪問期間與您的小組討論。

請您讓我們知道您的團體是否可以配合，我們將據此安排我們的訪問時間。在關於酒店和交通安排，我很感謝您的安排，但接下來我會交由我的助理黛比處理，因為我們可能會在該地區的同時訪問其他幾家公司。

我們期待著與您在 11 月的會議。

您忠誠的，

Aden Chen

銷售主管

咖啡共和國

達人提點 II

這篇範例說明了當受到邀約時的回函方式。在美語系國家裡，明確的 Yes or No 一般來說都會在文章的一開頭說明。如同範例，在回函的開頭便明確的表示很榮幸受邀參加。當然若是無法參加，也是同樣在文章開頭說明，之後再說明無法參加的原因。

說明確定會參加之後，可以在文章的中段裡提出自己在與會時的要求。在範例中提到的是私人會議的要求。當然，提出自己的要求同時也應該說明要求的原因，如文中所提到"由於政府更改進口法，所以可能會引起貨物上的一些問題，所以需要進行討論"的這個理由。

在回函的最後，由於邀請函有提及交通及飯店的預約，所以我們必須告知對方我們是否需要幫助，並禮貌地跟對方說期待與對方見面。

A 字彙

1. **Globally**: (adv) 全世界範圍地

   SARS affects the economy globally.

   SARS 對經濟的影響遍及全球。

2. **Region**: (n) 地區，地帶；行政區域

   It will become a long distance call once you cross region.

   只要你跨出行政區域則會變成長途電話。

3. **Trend**: (n) 趨勢，傾向；時尚

   The current trend is towards casual sporty wear.

   目前的趨勢是穿著輕便的運動服裝。

4. **Effect**: (n) 造成；產生；招致

   Careless driving might have a serious effect on road safety.

   粗心駕駛對道路安全有嚴重影響。

5. **Arrange**: (v) 安排；籌備

   John arranged a surprise party for his parent's 30th anniversary as a present.

   John 替他的父母安排了一個 30 週年的驚喜派對當作禮物。

6. **Supply**: (n) 供給，供應

   Fewer people are in the supply of small parts for replacement.

   越來越少人提供備用小零件。

7. **Lead-time**: (n) 生產時間；從訂貨到交貨的時間；交期

Lead-time often is an important factor for buyers to make decision where to place order to.

交期通常是採購在決定採買處的一個重要因素。

8. **Accordingly**: (adv) 照著；相應地

Students should obey the school rules accordingly.

學生應該照著校規走。

Unit 2 商業旅行 Business Travel

2-1 年度考核邀請函 Annual Audit Invitation

Excelics Microwave Electronics, Inc.
Cor.3rd.St., 3rd Ave., MEPZA, Lapu-Lapu City,
6015 Manila, Philippines

Mr. Robert Rosenberg,
Suite 110, Augustine Drive,
Santa Clara, CA 95054

Dear Mr. Rosenberg,

We are pleased to welcome you to visit our **facility** for the **annual** audit.

Ms. Emily Chou **informed** me that you will arrive in Manila on flight CX907 on Wednesday February 11th, at 09:50 am. Your personal **assistant** during your visit Ms. Alanna Wells will meet you at the airport and help you with all your travel needs. If there are any special **requirements** during your visit, please do not hesitate to let her know.

In regard to the hotel and restaurant **reservations**, please let me know the location and cuisine **preferences**, and I will **arrange** accordingly. I look forward to meeting with you.

Sincerely,
Damien Young, General Manager

中譯

親愛的 Rosenberg 先生，

我們很高興地歡迎您來參觀我們的工廠及進行年度審核。

Emily Chou 女士告訴我，您將會搭乘 CX907 班機，於二月十一日週三早上九點五十分抵達馬尼拉。您的個人秘書 Alanna Wells 女士將在機場與您碰面，並為您所有的旅行需求提供協助。如果您訪問期間有任何特殊要求，請不要猶豫，讓她知道。

在關於酒店和餐廳預訂，請讓我知道您對地點和美食的喜好，我會做適當的安排。期待與您見面。

誠摯的

Damien Young,

總經理

達人提點 I

這篇邀請函是在已經確定客戶會來訪後再寄出的邀請（確認行程）函。

因此，此篇的重點是在確認來訪客戶的行程，並確認客戶來訪時的需求。

文章的開頭應該先表示歡迎來訪之意，並說明來訪的目的。接下來即是確認來訪的時間及方式，並說明相互連絡的方式。注意來訪的日期、時間及接待人員的確認是非常重要的。

信件的最後應詢問對方是否有任何特殊需求，以便幫對方做好事前準備，這樣可以讓對方感受到歡迎之意。

A 字彙

1. **Facility**: (n) 場所

   This facility is built for kids with special needs.

   這個場所是為有特殊需求的小孩所設立的。

2. **Annual**: (adj) 一年的；一年一次的

   Employees are entitled to an annual paid with leave of fifteen days.

   職員一年可享受十五天帶薪的假期。

3. **Inform**: (v) 通知，告知，報告

   We were just informed that the flight will be delayed.

   我們剛剛被告知飛機將會延誤。

4. **Assistant**: (n) 助手，助理；助教；店員

   Kelly is interviewing as the assistant to the CEO.

   Kelly 正在應徵當ＣＥＯ的助理。

5. **Requirement**: (n) 要求；必要條件；規定

   Amber is trying her best to fill the requirements for promotion.

   Amber 正盡她所能得準備晉升的一切條件。

6. **Reservation**: (n) 預訂；預訂的房間或席座

   It is highly unlikely to get a table at the trendy restaurant in New York without reservation.

   在紐約，想要在沒有預約的情況下於一間流行的餐廳拿到位子幾乎是不可能的事。

7. **Preference**: (n) 偏愛的事物（或人）

Which is your preference, tea or coffee ?

你喜歡喝哪一樣，茶還是咖啡？

8. **Arrange**: (v) 安排；籌備

His personal assistant always arranges his travel schedule.

他的個人助理總是替他安排他的旅遊行程。

2-2 邀請回覆 Invitation Response

例文

Date: January 31st, 2015
From: rrosenberg@remecbb.com
To: d_young@excelics.com
Subject: Vendor facility audit and itinerary confirmation

Dear Mr. Young,

I appreciate your **proper** response in **regard** to my visit to your facility next month.

I will arrive in Manila on flight CX907 on February 11th **indeed**. About the hotel reservations, any location downtown with breakfast and high-speed internet is **preferred.** Dinner reservation is not **necessary**. I think it will be a good chance for me to **explore** the city during this trip.

I **look forward to** meeting with you and your staffs at your facility **as well**.

Sincerely,
Robert Rosenberg, Vendor Management

中譯

日期：2015 年 1 月 31 日
從：rrosenberg@remecbb.com
給：d_young@excelics.com
主旨：供應商設施考核和行程確認

親愛的 Young 先生，

很感謝您適切回應我下個月至貴公司的訪問。

我確實會在二月十一日搭乘 CX907 班機抵達馬尼拉。關於酒店的訂房，任何位於市中心並提供早餐和高速互聯網的酒店將為首選。晚餐的預訂是沒有必要的。我認為這對我來說將是一個遊覽城市的好機會。

我期待與您和您的員工在您的工廠見面。

誠摯的
Robert Rosenberg
供應商管理

在回覆確認用的邀請函時，直接、簡潔的回答是最簡單，也是最受歡迎的回函方式。

如同範例裡說明的一樣，日期、時間及地點的確認不但不可或缺，在回函裡不單單只是回答"正確，Correct"就好，而是要詳細說明行程，這也是比較好的答案。

關於其它細項的確認，亞州人往往會因為"不好意思"而回答不需要幫忙。這跟歐美國家的回答方式是背道而馳的。當回覆這樣的問題時，最好的答案便是明白地說明自己的需要及不需要。這樣不但方便對方作業，也可以讓自己得到最好的幫助。

最後，說明自己的感激當然還是必要的。

字彙

1. **Proper**: (adj) 適合的，適當的，恰當的
It is not proper to wear flip flops to a hotel.
穿人字拖進飯店是不恰當的。

2. **Regard**: (n) 關係，事項
In regard to that accident, it was not his fault.
關於那個意外並不是他的錯。

3. **Indeed**: (adv) 真正地，確實，實在
He is indeed a good friend.
他確實是一個好朋友。

4. **Preferred**: (adj) 更好的；被喜好的；優先的
A seat by the window is preferred.
窗邊的座位是比較好的。

5. **Necessary**: (adj) 必要的，必需的
A bicycle is necessary if you live in Santa Barbara.
如果你生活在勝塔巴巴拉，自行車是必要的。

6. **Explore**: (v) 探測；探勘；在……探險
She planned to explore the United States by driving across the country.
她計畫用開車跨越全美的方式探險美國。

Unit 3 談判 Negotiating

3-1 降價需求 Price Reduction Request

例 文

Nov. 1st, 2014

Mr. Robert Lin

Sales Director

Apollo Semiconductor, Inc.

Dear Mr. Lin,

Subject: Annual Pricing Review

Allow us to begin by thanking you for all your business support thru out the years. It has always been our pleasure to have Apollo Semiconductor as our **primary** business partner. However, I am sure you are aware of the **concern** about the drop in gold price internationally. Since gold price is the main **driven cost** of the chip production we purchase from you, we kindly ask you to review the annual selling price for 2015. As you might be aware, several local chip makers have offered us a lower price **solution**. The lowest discount we have been offered is close to 30% lower than what we are purchasing from Apollo now.

As much as we **honor** the business relationship between Apollo and us, the high production cost and price **reduction** pressure from customers are forcing us to consider other possible solutions if the price from Apollo is not **negotiable**.

We realize this will not make your situation easier. However, please kindly review this matter and **amend** the annual price for 2015 accordingly. We are hoping to get a minimum of 25% off the current price for all chips and change the payment term from Net 30 days to Net 60 days.

We look forward to receiving a report on your new pricing strategy soon.

Sincerely yours,
Oliver Chen
Global Supply Chain Manager

中譯

2014 年 11 月 1 日
Robert Lin 先生
銷售總監
阿波羅半導體公司

親愛的 Lin 先生，

主題：年度定價檢討

首先，讓我們感謝您多年來對本公司所有業務上的支持。一直以來，我們很高興有阿波羅半導體作為我們主要的貿易夥伴。然而，我相信您非常瞭解近日國際金價大幅下降的問題。由於黃金價格是我們向您購買芯片生產的主要驅動成本，我們懇請您檢閱 2015 年的年度售價。正如你所知的，一些本土芯片廠商已經為我們提供了一些較低的價格解決方案。最低的折扣是比我們從阿波羅現在購買的價格低近 30%。

雖然我們尊重阿波羅和我們之間的業務關係，但由於高生產成本及客戶端的價格下調壓力，如果阿波羅半導體無法降價，此正迫使我們考慮其他可能的解決方案。

我們知道這不是容易的狀況。但是，請您檢閱並修改 2015 年的年度價格。我們希望能得到至少 25%的折扣，並將付款方式由到貨後 30 天內付款改為 60 天。

我們期待儘快收到您新的定價策略。

您忠誠的，

Oliver Chen
全球供應鏈經理

達人提點 I

無論是提出什麼樣的談判要求，簡單的寒暄是絕對必要的。"首先，讓我們感謝您多年來對本公司所有業務上的支持"這就是一個制式化但不可缺的寒暄。切入主題時，說明背景原因是第一要件。如同範例，在文中先說明國際金價下跌，而產品的驅動成本為金價，因此寄件者認為有談判的空間，進而提出要求。提出要求的同時，我們如果有可以幫助說服對方的理由，則可以增加這方面的訊息在信件中。例文中就有提到其它廠商願意以更低的價錢販賣給作者。在這樣的信件中，給與適當的壓力也是不錯的做法。如例文中提到的"若無法降價，我們則不得不轉向其它廠商。"

最後，如同所有要求文一樣，具體要求一定要列出。例如：我們要求最少 25% 的降價，並延長付款天數。

A 字彙

1. **Primary**: (adj) 首要的，主要的

   A primary cause of their divorce is his busy job .

   他忙碌的工作是他們離婚的主要原因之一。

2. **Concern**: (v) 擔心，掛念；關懷

   Parents always concern about their children no matter how old they are.

   不管孩子幾歲，父母總是擔心他們。

3. **Driven cost**: 驅動成本

   The driven cost of a product often decides the selling price.

   一件物品的價錢往往取決於驅動成本。

4. **Solution**: (n) 解答；解決；解釋

   One of the solutions to global warming is to stop air pollution.

   全球暖化的解決方法之一是停止空氣污染。

5. **Honor**: (n) 榮譽；名譽，面子

   It is an honor to be invited to the party.

   獲邀參加這個派對是一種榮譽。

6. **Reduction**: (n) 減少；削減

   Reduction of household expenses gets harder when kids grow older.

   當孩子年紀越來越大，家用開支的削減就越來越困難。

7. **Negotiable**: (n) 可協商的

Price is often time negotiable while purchasing merchandise at the night market.

當在夜市買東西時，價錢往往是可協商的。

8. **Amend**: (v) 修訂，修改；訂正

The company rule was amended that employees no longer need to wear uniform.

公司的規定已修改，員工不再需要穿制服。

3-2 回絕降價要求並提供低價政策
Reject price reduction request but offer low cost solution

Nov. 3rd, 2014
Mr. Oliver Chen
Purchasing Head
Excellent Design, Inc.

Dear Mr. Chen,

Thank you for your letter of Nov. 1st in which you request a more **competitive** price for our chips. Please understand that we always appreciate your business and fully realize the importance of providing your company with quality products under **condition** that will **permit** you to win a **commending** share in your market. As you might consider the possibility of price drop due to the decrease in gold price, I am sorry to tell you the driven cost of our chip is actually the design cost and testing cost, instead of the **material** cost. Therefore, the offer we have been given to you is the best price we could give for the models you currently purchase.

However, we do appreciate the partnership between Excellent Design and Apollo Semiconductor; therefore, we would like to offer **alternative** solutions for all chips you currently purchase from us. The basic designs of the chips are exactly the same as the models you purchase now, but we **simplified** the testing processes. By doing so, we are able to offer you a 20% price cut for all chips. As you may be concerned about the quality, we are happy to provide free samples for you to do testing, and also provide the same **warranty** as always.

In regard to the payment term, I am sorry to tell you that the best we can offer is to change the payment condition from Net 30 to Net 45. Any condition longer than that will be difficult for us to run the business. I hope you can understand our position.

We look forward to hearing from you soon.

Sincerely yours,
Robert Lin,
Sales Director

中譯

2014 年 11 月 3 日
Oliver Chen 先生
採購主管
優秀設計公司

親愛的 Chen 先生，

感謝您 11 月 1 日來信為我們的芯片要求一個更有競爭力的價格。請您理解，我們一直很感謝您對本公司生意上的貢獻，本公司也瞭解提供貴公司優質的產品以允許您贏得您市場份額的重要性。雖然你可能會認為金價下跌可能影響賣價的下調，但我很遺憾地告訴您，本公司芯片的驅動成本實際上是設計成本及檢測成本，而不是材料成本。因此，目前我們所提供給您的報價已經是你目前購買該機型的最佳價格。

不過，我們也明白優秀設計和阿波羅半導體之間的夥伴關係；因此，我們希望可以提供給您目前從我們這裡購買所有芯片的替代方案。替代方案的芯片基本設計是完全一樣的，但我們簡化了測試過程。如此，我們可以對所有芯片提供 20%的降價空間。而您所關切的品質方面，我們樂意為您提供免費的樣品供您做測試，同時維持相同的保修一如既往。

關於付款條件，我很遺憾地告訴您，我們可以提供最好的付款條件為到貨後 30 天內付款更改為 45 天。任何其他條件將會讓我們的經營陷入困難。我希望您能理解我們的立場。

我們期待您的佳音。
您忠誠的，
Robert Lin

銷售總監

雖然說這封信件的目的是要拒絕對方的要求，但是開頭的感謝文還是必要的。先以感謝對方的文字表明你和對方是在同一陣線上。值得注意的是英語商用書信寫法與中文不同，除了表明感謝之外，往往會在開頭重複對方的要求，以免誤會。而後，在回絕對方時，應該要先說明為什麼對方的要求是不合理的。先讓對方瞭解理解錯誤的地方，進而告訴對方他的要求是不被接受的。

若有其它替代或解決方案時應於這時提出。解決方案應說明內容並解釋可以達到客戶要求的原因。例如在文章中所提到的替代方案之所以可以比原來的方案便宜是因為減少了測試。

替代方案難免會造成客戶的擔心，所以明白說明不同點以及提供風險資訊是必要的。

在文章的最後，建議還是再次說明自己與客戶是在同一立場。這可以幫助你的客戶相信你的產品，進而接受建議。

A 字彙

1. **Competitive**: (adj) 競爭的；經由競爭的，競爭性的
   A sales with competitive personality always has an outstanding accomplishment.
   很有競爭性格的業務通常有很好的業績。

2. **Condition**: (n) 條件
   He is allowed to leave school early on condition that he turns in homework on time tomorrow.
   他可以提早離開學校，條件是明天按時交功課。

3. **Premit**: (v) 允許，許可，准許
   His parents permitted him to go skying.
   他的父母允許他去滑雪。

4. **Commend**: (v) 稱讚；讚賞
   The mayor commended the police for his bravery.
   市長表揚了那個警察的英勇行為。

5. **Material**: (n) 材料，原料
   Sapphire is a common material for LED chips.
   藍寶石是ＬＥＤ晶片的常用原料。

6. **Alternative**: (adj) 替代的；供選擇的
   She is looking for an alternative route to go to work.
   她在找上班的替代路線。

7. **Simplify**: (v) 簡化，精簡；使單純；使平易

Chatting apps, such as Line simplified the communication between people.

像 LINE 這樣的聊天軟體簡易了人們之間的溝通。

8. **Warrenty**: (n) 保證書；保單

A 5-year 40,000 miles warrenty is a standard to most cars.

幾乎所有的車都有 5 年 4 萬英里的保證。

Unit 4 推銷 Sales

4-1 投資討論 Invent Invitation

March 27th, 2015
Ms. Jennifer Azusa

Dear Ms. Azusa,

It was my great pleasure to write this letter to you. I would like to thank you for giving me such a great experience to have the **authentic** taste of the Hawaiian pancake here in Honolulu. As a food lover who almost tasted everything around the world, I would say the Whippy Pancake is something I have never tasted before. I believe this authentic flavor of Hawaii can be **expended** to the world in a very **popular** way.

With that said, I am asking your **permission** to allow my team to bring this wonderful flavor back to Taiwan and the permission to open the Azusa Pancake in Taipei, Taiwan. As one of the fastest **economic** growth cities in the world, the demand for quality food and desserts is increasing. Recent data also shows that over 70% of the people enjoy the morning coffee and western breakfast during days off. Although, we still find that quality authentic western breakfast is still **lacking**, it is a big market for us to explore.

Our basic plan is to open the first store at the **financial district**, XinYi District, Taipei. We believe it can generate about 100 thousand USD of revenue per month at the first 6 months, and looking at a 30% growth per year by expanding new stores and increasing sales items.

In regard to the share holding, by law, foreign firms can not hold more than 50% of the **capital**. Therefore, we would like to offer 49% of the share capital to Azasa Pancake, and NW Capital holds the rest of 51%. We believe that this is a fair offer and hope Azasa Pancake can consider working with us and let people in Taiwan have the chance to enjoy the wonderful flavor created by you.

I sincerely hope I can get a chance to arrange a meeting with you and present my business proposal in person.

Sincerely yours,
Neil Wu
General Manager
FirstCapital, Corp.

中譯

2015 年 3 月 27 日
Jennifer Azasa 女士

尊敬的 Azusa 女士，

很高興寫這封信給您。我要感謝您給了我這樣一個美好的經驗，讓我有機會體會到在夏威夷檀香山的正宗夏威夷鬆餅的味道。作為一個美食愛好者，幾乎嚐遍世界各地美食的我，Whippy Pancake 是我以前從來沒有吃過的。我相信夏威夷這個道地的風味，可以成為世界的一種流行。

因此，我希望您能同意讓我的團隊把這個美好的味道帶回台灣，讓我們在台灣台北開設 Azusa 鬆餅。台北作為世界上經濟增長最快的城市之一，優質食品和甜點的需求正在增加。最近的數據也表明，70%的人喜歡在休息日的早晨享受咖啡和西式早餐。雖然，我們也發現，正宗的西式早餐仍然不足。因此這是一個很大的市場，值得我們去探索。

我們的基本計劃是在台北的金融區，信義區，開設第一家專賣店。我們相信它可以在 6 個月內產生約每月 10 萬美元的收入，並期待藉由開設新店面及增加銷售品項來期待每年 30%的成長。

有關持股方面，根據法律規定，境外公司不得持有 50%以上的資本。因此，我們願意提供 49% 的股本給 Azasa 鬆餅，NW 投信將持有剩下的 51%。我們相信這是一個公平的提案，希望 Azasa 鬆餅可以考慮與我們合作，讓台灣民眾有機會享受由您創造的美妙味道。

我衷心希望我能有機會安排與您會面，並提出我的商業計劃書。

您忠誠的，

Neil Wu

總經理

NW Capital

達人提點 I

這是一個推銷自己使對方願意與你合作的例文。

首先，你必須告訴對方你為什麼喜歡對方。這裡的對方可以指公司、產品，甚至個人。明確地告訴對方他的優點在哪裡，想要和他合作的原因為何。

接下來，就是推銷自己的時候。而這裡的自己，同樣的也是可以指自己個人、公司、地點等等重要的因素。先讓對方瞭解自己，讓對方有充分的資訊以瞭解自己是跟什麼樣的人、事、物打交道，以決定要不要與你合作。

之後則是說明自己所想的生意計畫，讓對方有時間模擬與你合作的好處與壞處。持股的計畫也可以在同時間提出。

A 字彙

1. **Authentic**: (adj) 真正的，非假冒的

The authentic drawing is now displaying in the city museum.

這幅畫的真蹟目前在城市博物館展覽。

2. **Expended**: (n) 消費，花費

She expended all her youth on her boyfriend who cheats.

她將她的青春都花費在她會出軌的男友身上。

3. **Popular**: (adj) 普通的，廣為流傳的，流行的

Cola is a popular drink in America.

在美國，可樂是個廣為流行的飲料。

4. **Permission**: (n) 允許，許可，同意

You cannot go out without mom's permission.

你不能沒經過媽媽的同意就外出。

5. **Economic**: (adj) 經濟上的；經濟學的

Many students can only go to public schools for economic reasons.

出於經濟上的原因許多學生只能上公立學校。

6. **Lack**: (v) 缺少；沒有

It is fortunate to feel like you don't seem to lack anything in life.

感覺你的生命中什麼都不缺是很幸福的。

7. **Financial district**: 金融區

   It is his dream to work at the largest financial district of the world, New York.

   在世界上最大的金融區紐約工作是他的夢想。

8. **Capital**: (n) 資本；本錢

   The company is running out of capital, so it is filing for bankruptcy.

   這間公司花完資本所以已經申請倒閉。

4-2 邀請回覆
Invitation Response

March 30th, 2015
Mr. Neil Wu
General Manager
NW Capital, Corp.

Dear Mr. Wu,

Thank you for your letter of September 27th in which you mention an interest in opening up Azusa Pancake in Taipei Taiwan.

As you know that Azusa Pancake has been a local business in Honolulu for over 30 years. Our **founder** Anna Azusa opened up the restaurant in her backyard, and her business **philosophy** was not about making a **fortune**, but to make the healthy and tasty breakfast for the neighborhood. We are lucky that we have been loved by all our customers for so long, and now have the chance to expand our homey flavor to other places as well. Although we have several concerns about **maintaining** our business philosophy and our local flavor in the busy financial districts as Taipei, we are open for **discussion** and look forward to hearing your **proposal**.

Please let us know when you would like to set up the meeting.

It will be very much appreciated if we can receive **additional** information about your company and your business proposal before we meet. This would serve to make such a meeting more **effective**.

We look forward to hearing from you.

Best Regards,
Jennifer Azusa
General Manager
Azusa Pancake

中譯

2015 年 3 月 30 日

Neil Wu 先生

總經理

NW 投信公司

尊敬的 Wu 先生，

感謝您在 9 月 27 日提到有興趣在台灣台北開設 Azusa 鬆餅的來信。

正如你知道 Azusa 鬆餅已經在檀香山當地有超過 30 年的歷史。我們的創始人 Anna Azusa 在自家後院開闢了這家餐廳，她的經營理念不是賺錢，而是為了讓鄰居能夠品嚐到健康美味的早餐。我們很幸運，我們一直長期受我們所有的客戶喜愛，現在還有機會擴大我們的溫馨家庭氣息到其他地方。雖然我們對在繁忙的台北金融區保持我們的經營理念有一些疑念，但我們願意開放與討論，並期待著聽到您的建議。

請讓我們知道您想設置會議的時間。且如果我們能在與您見面前收到有關您公司的訊息及您的商業計劃書，將有助於使會議更有效。

我們期待您的回音。

最好的問候，

Jennifer Azusa

總經理

Azusa 鬆餅

達人提點 II

在回信時我們會把收到信件的日期列上，並表明書信的內容。

接下來無論接受邀約與否，都應當先表明自己的立場及想法。如同在例文中所提到的商家歷史及創辦的理念，還有對未來的展望以及對提案的疑問。在例文中的結論是商家願意開放討論，所以在最後一個段落裡，我們可以提出希望在下次會議時所希望看到的文件及希望收到的相關訊息。而如果你的答案是否定的，則建議在後段裡提及回絕的理由，並對於提案表示感謝。當然，若可以同時提出其它可行方案，例如建議詢問其它廠商等，則非常建議在此提出。

A 字彙

1. **Founder**: (n) 創立者；奠基者；締造者
   The founder of a company normally has great creativity.
   一間公司的創立者通常有很好的創造力。

2. **Philosophy**: (n) 哲學，人生觀
   "Work hard play hard" is his philosophy.
   「認真工作用力玩」是他的人生觀。

3. **Fortune**: (n) 財產，財富；巨款
   He received a large fortune when winning the lottery.
   他在贏得樂透時得到了一大筆財富。

4. **Maintain**: (v) 維持；保持；使繼續
   Earning a fortune is easy, but maintaining it is hard.
   賺取一筆財富是容易的，但維持它是困難的。

5. **Discussion**: (n) 討論，商討；談論
   We had a discussion on next year's annual goal.
   我們就明年的年度目標進行了討論。

6. **Proposal**: (n) 建議，提議；計畫；提案
   She just filed in a proposal to build the online banking system.
   她剛剛提出架構網上銀行系統的提案。

7. **Additional**: (adj) 添加的；附加的；額外的

The professor required additional information to support the theory.

教授要求提出額外的資訊以支持這個理論。

8. **Effective**: (adj) 有效的

Penalties are effective in keeping out speeding.

罰金能有效防止超速。

Unit 5 協定 Agreement

5-1 保密協定 Mutual Non-Disclosure Agreement

THIS **MUTUAL NONDISCLOSURE** AGREEMENT is made and entered into as of August 16th, 2010 between ABC Corp. with offices at Taipei, Taiwan and Alpha Inc., with offices at Tokyo, Japan.

Purpose. The parties wish to explore a business opportunity of mutual interest and in connection with this opportunity, and each party may disclose to the other party certain **confidential** technical and business information which the disclosing party desires the receiving party to treat as confidential.

"Confidential Information" means any information disclosed by either party to the other party, either directly or indirectly, in writing, orally or by inspection of tangible objects, including without limitation documents, prototypes, samples, plant and equipment, research, product plans, products, services, customer lists, software, developments, inventions, processes, designs, drawings, engineering, hardware configuration, marketing materials or finances, which is designated as "Confidential," "**Proprietary**" or some similar designation.

Confidential Information shall not; however, include any information which (i) was publicly known and made generally available in the public **domain** prior to the time of disclosure by the disclosing party; (ii) becomes publicly known and made generally available after disclosure by the disclosing party to the receiving party through no action or inaction of the receiving party; or (iii) is required by law to be disclosed by the receiving party.

Non-use and Non-disclosure. Each party shall not use the Confidential Information of the other party for any purpose except to **evaluate** and engage in discussions concerning a potential business relationship between the parties. Neither party shall disclose any Confidential Information of the other party to third parties.

Return of Materials. All documents and other **tangible** objects **containing** or representing Confidential Information which have been disclosed by either party to the other party, and all copies thereof which are in the possession of the other party, shall be and remain the property of the disclosing party and shall be promptly returned to the disclosing party upon the disclosing party's written request.

Miscellaneous. This Agreement shall be governed by the laws of Taiwan. This Agreement may not be amended, nor any obligation waived, except by a writing signed by both parties hereto.

ABC Corp.		Alpha Inc.	
By:	Signature	By:	Signature
Name:	CK, Tseng	Name:	Takeshi, Fujta
Title:	Vice President	Title:	General Manager
Date:	August 1st. 2010	Date:	August 5th, 2010

中譯

此為 2010 年 8 月 16 日位於台灣台北的ＡＢＣ公司與位於日本東京的 Alpha 公司所訂的相互保密協議。

目的 - 雙方希望探討共同感興趣的商業機會，因此，每一方可以披露某些機密技術和商業訊息，但對方須以保密文件對待。

“機密信息”是指由任何一方以直接或間接的方式披露給對方的任何信息，包括書面、口頭或通過檢查的有形物體，包括但不限於被指定為“機密”，或“專有”的文檔、原型、樣品、廠房及設備的研究、產品計劃、產品、服務、客戶名單、軟體、開發、發明、工藝、設計、圖紙、工程、硬件配置、營銷材料或資金。但機密信息不得包括（一）眾所周知，並已披露在公共領域的任何信息；（二）未經接收方要求，由披露方主動提供並成為眾所皆知的信息；或（iii）按法律規定必須由接收方披露。

不使用和不披露。任一方除了評估和從事有關各方之間的潛在業務關係的討論外不得使用對方的保密信息於任何目的。任何一方不得透露給第三方任何機密信息。

文件繳回。所有已被披露給對方的文件和其他實物或包含代表機密的信息，包含並且拷貝文件應當保持為披露方的財產。

其他。本協議受台灣法律管轄。本協議除非由雙方同意，否則不得修改，或得以任何豁免。

ABC Corp.	Alpha Inc.
簽名：簽名	簽名：簽名

姓名:	CK, Tseng	姓名:	Takeshi, Fujta
職稱:	副董事長	職稱:	執行長
日期:	2010 年 8 月 1 日	日期:	2010 年 8 月 5 日

達人提點 I

企業間的保密協定基本上都有一個既定的書寫方式。各間公司大同小異。

契約制定日期、雙方公司的正式名稱，以及公司的地點甚至地址都需要在協定的一開頭明白列出。

第二段則是說明這個保密協定的目的。目的也通常是寫出因雙方需就某些原因交換商業機密，因此需要這個保密協定來約束雙方不將情報外露。

在接下來的內文部份則應當說明保密協定的內容。什麼樣的情報、內容、物品等是被規例為被保密的內容。同時也應列出相互對保密內容的處理方式，例如如何將情報繳回或銷毀。

最後，則是一定會列出法律追溯的地點。通則是以寄出保密協定的那一方的地方政府機關為主。

字彙

1. **Mutual**: (adj) 相互的，彼此的

   It is a mutual understanding that the car belongs to the father.

   我們彼此都瞭解這台車是屬於父親的。

2. **Nondisclosure**: (n) 不被透露；非公開

   This is a nondisclosure lawsuit.

   這是非公開訴訟。

3. **Confidential**: (adj) 祕密的；機密的

   All documents listed as confidential cannot be brought out of this building.

   所有被列為機密的文件都不能被帶出這個建築物。

4. **Proprietary**: (adj) 專賣的；專利的

   Steve owns a proprietary cleaner which made him a fortune.

   Steve 擁有一種專利洗劑讓他賺了一大筆錢。

5. **Domain**: (n) 領域，範圍

   Finance is in my domain.

   金融是他的領域。

6. **Evaluate**: (v) 估價，評價；估算

   It is difficult to evaluate your own ability.

   評估自己的能力是很困難的。

5-2 與員工間的保密協定 Employee Confidentiality Agreement

This Agreement is made between Giana Camarena ("EMPLOYEE") and Matsumoto Inc. ("Matsumoto"), on April 1st, 2013.

EMPLOYEE will perform services for Matsumoto, which may require Matsumoto to disclose confidential and proprietary information ("Confidential Information") to EMPLOYEE. Accordingly, to protect the Matsumoto Confidential Information that will be disclosed to EMPLOYEE, the EMPLOYEE agrees as follows.

A. EMPLOYEE will hold the Confidential Information received from Matsumoto in **strict** confidence and shall exercise a reasonable degree of care to **prevent** disclosure to others.

B. EMPLOYEE will not disclose or **divulge** either directly or indirectly the Confidential Information to others unless first **authorized** to do so in writing by Matsumoto.

C. EMPLOYEE will not **reproduce** the Confidential Information nor use this information **commercially** or for any purpose other than the performance of his/her duties for Matsumoto.

D.Matsumoto reserves the right to take **disciplinary** action, up to and including termination for **violations** of this agreement.

Signing below signifies that the EMPLOYEE agrees to the terms and conditions of the agreement stated above.

Matsumoto Inc.	EMPLOYEE
______________________	______________________
Human Resource Signature	Employee Signature
Date: ______________	Date: ______________

中譯

該協議是由 Giana Camarena（“員工”）和松本公司（“松本”）在 2013 年 4 月 1 日所簽訂的協議。

在員工替松本進行服務時可能需要松本透露相關機密和專有信息（“保密信息”）給員工。因此，為了保護松本將要披露的員工保密信息，員工同意如下。

A. 員工將嚴格保密於松本收到的機密信息，並合理謹慎，以防止洩露給他人。

B. 員工在未經松本書面授權前，不會透露或洩露任何直接或間接的機密信息給他人。

C. 員工在他 / 她於松本的職責之外不會拷貝或使用這些信息作為商業用途。

D. 本保留採取紀律處分，直至並包括終止對違反本協議的權利。

下面簽名表示該員工同意上述協議的條款和條件。

松本公司	僱員
____________________	____________________
人事部簽章	僱員簽章
日期 ____________	日期 ____________

達人提點 II

同樣為保密協定，這篇例文的對象不同，是員工與公司間的協定。寫法較為輕鬆，但大同小異。在文章的開端，我們會列出對象，即是員工的正式名稱和公司在哪一天（日期）所設的保密協定。並且為了這類書信的通用性，我們會在員工及公司的名稱後（以下稱謂）放入，這麼做的目的是同一封信函可以同時被利用在其它員工身上，只需要更改開頭的正式名稱即可。這些都是制式化的寫法。

再來則是寫出需要這份保密協定的原因，並將內容以條例是的方法列出。最後再由雙方簽名，並寫上日期便完成了這份協定。

A 字彙

1. **Strict**: (adj) 嚴謹的，精確的

   The school has strict rules to control students behavior.

   學校有很嚴謹的規則以控制學生的行為。

2. **Prevent**: (v) 防止，預防

   Apples can prevent sickness .

   蘋果能預防生病。

3. **Divulge**: (v) 洩露；暴露

   If any confidential information is divulged, you will be fired.

   若任何機密的情報遭洩露，你將會被革職。

4. **Authorized**: (adj) 經授權的；經批准的

   Only US citizens are authorized to enter the military base.

   只有美國公民有經授權可以進入軍事基地。

5. **Reproduce**: (v) 複製；翻拍；複寫；再上演

   Grandmother was very touched when we reproduced the picture of her and grandfather.

   當我們翻拍外婆與外公的相片時，外婆非常的感動。

6. **Commercially**: (adv) 商業上；通商上

   We are not allowed to do any activities commercially when we are off duty.

   當我們不在工作時，我們不被允許做任何商業行為。

7.**Disciplinary**: (adj) 紀律的；懲戒的

I prefer not to take the disciplinary action if you tell the truth.

如果你說實話，我傾向不使用紀律處分。

8. **Violation**: (n) 違反行為

Any departure from this order is considered a violation of the laws .

任何背棄這個命令的行為都被認為是違法行為。

6-1 基礎市場分析師招募文
Entry Level Market Analyst Recruiting

Entry Level Marketing

Proweb Design & Marketing, Corp.

Proweb Design & Marketing Corp. helps our clients to manage the online marketing and web **visibility** in medical industries professionally. Starting from branding, website design to online marketing and web visibility analysis, Proweb & Marketing is your one stop solution.

We're looking for an **energetic**, self-motivated person to help with the **creation** of marketing and email **campaigns**. If you are a quick-learner, good with numbers and detail-oriented, you might be who we're looking for. This full-time position is a career opportunity to learn about marketing at a fast-growing online company, and works with an awesome group of people.

Essential Duties and Responsibilities include:

Digital and Traditional Marketing:

Email marketing: draft emails, manage send lists and schedule; track responses and performance; analyze results and **generate** reports

Websites: perform heavy repetitive website testing, report issues, and suggest improvements; write requirements and supporting documentation

Content Marketing: write content for websites, emails, newsletters, social media, blogs, etc. Help develop and update sales collateral and marketing materials

Research & Analysis, Reports, & Presentations:

Conduct competitive analysis, including heavy research on competitors' activities, services and pricing; collect competitors' marketing materials.

Advanced usage of MS Excel and PowerPoint to produce reports. Assist with the development of the annual marketing plans and budgets. Manage email send and snail mail lists.

Location: XinYi District, Taipei, Taiwan

初階市場行營人員

Proweb 設計與市場營銷公司

Proweb 設計與營銷公司幫助我們的客戶以專業的方式管理在醫療行業的網絡營銷和網站的知名度。從品牌開始，網頁設計、網絡營銷和網絡可視性分析，Proweb 和營銷是您的一站式解決方案。

我們正在尋找一個充滿活力的、有上進心的人，以幫助建立營銷和電子郵件活動。如果你是一個學習快速、善於用數字和注重細節的人，你可能會是我們要找的人才。

這是一個想要在快速增長的網絡公司學習如何經營，以及與一群很棒的團隊工作。

重要的工作職責包括：

數位和傳統營銷：

●電子郵件營銷：編寫電子郵件草稿，管理發送清單和時間表，追蹤回覆及效果，分析結果並製作報告

●網站：執行繁重的重複性網站測試，提出問題報告，並提出改進意見，編寫要求及輔助文件

●內容編寫：撰寫網站、電子郵件、通訊、社交媒體、部落格等內容

●幫助制定和更新銷售資料和營銷材料

研究與分析，報告與發表：

●競爭者分析，包括對競爭對手的活動、服務和定價的多種研究；蒐集競爭對手的營銷材料。

●高級程度的 MS Excel 及 PowerPoint 使用以製作報告。

●協助年度營銷計劃和預算的制定。

●管理電子郵件的發送和蝸牛郵件列表。

地點：台灣，台北，信義區

達人提點 I

首先，在這類型的廣告中最重要的就是告訴別人“我們要的是什麼”。無可厚非，第一點就是說明我們在應徵的是什麼工作。

接下來，則是自家公司的說明，告訴有興趣者自家公司的特色以及工作內容，並簡單說明想要找的是怎麼樣的人才。

詳細的工作內容及負責業務內容則應用條例的方式一一列舉說明。在此，我們不建議使用文章式的寫法，因為容易因過度累贅的文章內容反而造成詞不達意或說明不清的狀況。

無論寫任何關於“條件”或“條例”性的文章，我們都建議使用單點列舉式的方式撰寫。

上班地址及連絡方式也要記得在文章的最後補上。這看似理所當然的部份，其實卻是經常被遺忘的。

A 字彙

1. **Visibility**: (n) 能見度；明顯性；視程

   Drivers need to pay extra attention during raining days since the visibility is bad.

   雨天的能見度很差，駕駛應當更加小心。

2. **Energetic**: (adj) 精力旺盛的；精神飽滿的

   Babies are always energetic.

   嬰兒總是精力旺盛。

3. **Creation**: (n) 創造；創作；創立

   The landmark is the mayor's creation.

   這個地標是市長的創作。

4. **Campaign**: (n) 運動，活動

   Students are showing their incentive to join a campaign against smoking.

   學生們積極地參與一項禁煙運動。

5. **Essential**: (adj) 必要的，不可缺的

   Teacher's essential duty is to teach students how to be a good person.

   老師的必要責任是教導學生成為一個好人。

6. **Generate**: (v) 產生，發生

   Heat can generate power.

   熱可以產生能量。

7. **Content**: (n) 內容，要旨

Before purchasing a book, you'd better to check the content first.

在買書之前，你最好是先確認內容。

8. **Conduct**: (v) 實施；處理；經營，管理

They hired agents to conduct their affairs.

他們僱請代理人來處理他們的事務。

6-2 就業推薦函
Recommendation Letter

Ms. Emily Birds
101 Santa Clara Blvd,
Santa Clara, CA 94132

Dear Ms. Birds,

This letter concerns Bob Wang who I had the pleasure of working with during his college years as a part time temp in our department. Mr. Wang was working as not only my personal assistant but also the front desk staff. Therefore, I am in a position to offer some **concrete** information on his **particular capabilities**.

During his 2-year temp job as our staff here in ELS department, Mr. Wang **adapted** well to the pace of life here and got on extremely well with the people he came in contact with. I know for a fact that he was able to **establish** close relations with many members of our staff and our students. For two years straight, he was voted as "the best staff of the semester" by our teachers and students. He was not only a friendly and easygoing person to work with, but also was detail oriented and full of creativity.

The **caliber** of work he did for me was outstanding. For

2 years, he was responsible for all new students **enrollments**, international students housing, event planning, and advertisement. Often time, Mr. Wang was able to nail out the existing problems or points that needed improvements and gave good solutions or suggestions. He continually went extra-mile and **dutifully** met time limits. Working with him was in many ways a refreshing experience.

I can therefore recommend him without reservation.

Sincerely your,
Catharine Boyd
Director of ELS Dept.
Santa Clara University

中譯

尊敬的 Birds 女士，

這封信和 Bob Wang 有關。在他大學兼差時，我曾有幸與他在我部門一同工作過。王先生不僅是我的私人助理，也是前台的工作人員。因此，我適合提供您一些對於他特殊能力的具體信息。

在他擔任我們ＥＳＬ部門人員的兩年內，王先生非常適應這裡的生活節奏並與他接觸的人都維持很好的關係。我知道他能夠與我們的員工、學生及許多成員建立良好的密切關係。他連續兩年皆被我們教師和學生票選為“本學期最好的員工”。他不僅是一個友好、隨和的工作夥伴，也注重細節，且充滿創意。

他的工作表現非常出色。兩年中，他負責所有新生入學、國際學生住房、活動策劃和廣告。王先生總是可以指出需要改進的既有問題，並提供很好的解決方案或建議。他也不斷地突破，盡職盡責地在時間限制內完成工作。與他一起工作在許多方面是令人耳目一新的體驗。

因此，我可以毫無保留地推薦他。

真誠的，

Catharine Boyed

ELS 部總監

聖克拉拉大學

在撰寫推薦函時，為了要增加說服力，當然在文章的一開始我們要說明我們與被推薦者之間的關係，例如：什麼時候在什麼樣的狀況之下一同工作過。

接下來則是在內文中表明信的目的，加強說明與工作的關係，並列舉被推薦者的工作內容及可被推薦的原因。推薦者應當瞭解被推薦者需要這封信件的原因，且加以將被推薦者的優點與應徵工作的內容做連結，這樣才能有效地說服收信者。

A 字彙

1. **Concrete**: (adj) 具體的

   The boss requires a more concrete business plan.

   老闆要求一個更具體的商業計畫。

2. **Particular**: (adj) 特殊的；特定的；特別的

   This student requires particular care.

   這個學生需要特別的關心。

3. **Capability**: (n) 能力，才能

   He has the capability of long distance swimming .

   他具有長程游泳的能力。

4. **Adapt**: (v) 使適應，使適合

   As a foreign student, he tried hard to adapt himself to the new environment .

   當一個留學生，他努力使自己適應新的環境。

5. **Establish**: (v) 建立；設立；創辦

   Our school is full of history. It was established in 1890 .

   我們的學校建於 1890 年，歷史悠久。

6. **Caliber**: (n) 才幹

   His caliber of work is well-known by everyone in the company.

   他的工作才幹是公司裡的人都知道的。

7. **Enrollment**: (n) 登記；入會

The enrollment for the conference ends today at 5pm.

這個會議的登記將在今天下午五點結束。

8. **Dutifully**: (adv) 忠實地；忠貞地；盡職地

As a personal assistant of the CEO, she works dutifully and gains trusts from her boss.

作為ＣＥＯ的個人助理，她很盡職的工作並贏得老闆的信任。

Unit 7 通知 Notification

7-1 商品招回及退費通知案例一 Products recall and refund notification

Organic ForYou Issues **Voluntary** Allergy Alert on **Undeclared** Allergens in 100% Natural Blueberry Cream Spread

FOR IMMEDIATE RELEASE – June 20th, 2012 – Organic ForYou Inc. of Santa Clara, California is voluntarily **recalling** the 100% Natural Blueberry Cream Spread sold in the refrigerated section of all local grocery stores with "Use By 12/13/14" date, because it may not list peanuts in the ingredients. People who have an allergy or severe sensitivity to peanuts run the risk of serious or life-threatening allergic reactions if they **consume** this product.

The recall affects only 100% Natural Blueberry Cream **bearing** the "USE BY 12/13/14" date packaged in 20 oz. tray. The

Use By date is printed on the side of the sleeve. The 500 trays possibly affected were **distributed** to local grocery stores located in Santa Clara, California only. No other lots, products or stores are affected.

The voluntary recall was **initiated** by OrganicForYou after several consumers discovered that affected packages bearing the label for 100% Natural Blueberry Cream Spread **mistakenly** contained ingredients with peanuts.

Customers who have purchased the 100% Natural Blueberry Cream Spread may return it to your local store for a full refund. Customers with questions may contact OrganicForYou at 408-386-6732 from 9 AM-5:30 PM PST, Monday-Friday.

中譯

產品召回和退款的通知

OrganicForYou 未申報的 100% 天然藍莓奶油塗醬自願性過敏原通知
即時發布 - 2012 年 6 月 20 日 - 因為可能未列出成分花生，加州聖塔克拉拉 OrganicForYou 公司，加利福尼亞州自願召回所有在本地雜貨店所售出在冷凍區標示"使用期限為 12/13/14"的 100%純天然藍莓奶油塗醬。若對於花生有過敏或嚴重敏感反應的民眾消費此產品，此產品將會有造成嚴重，甚是危及生命的過敏反應的風險。

此次召回只影響到印有"使用期限為 12/13/14"的 100%純天然藍莓奶油塗醬，包裝為 20 盎司的托盤。使用期限被打印在套筒的一側。這 500 盤商品可能被分配到位於聖克拉拉當地各處的雜貨店，並只限於加利福尼亞州。沒有其他地段、產品或商店受到影響。

此召回是由幾位消費者發現 100% 純天然藍莓奶油塗醬誤添了花生，由 OrganicForYou 主動召回。

已購買 100%純天然藍莓奶油塗醬的消費者可將其退回給當地商店並獲取全額退款。如有疑問，可在西岸時間星期一至星期五，上午 9 時至下午 5 時連絡 (408)386-6732 OrganicForYou。

達人提點 I

產品召回的通知一般有分兩種，依照商品的不同做區分。一種是如同例文，商品為一般大眾化產品，銷售量多，因此招收公告只能在報章雜誌及網路上公告。撰文的方式較類似於新聞稿。第二種則是如汽車等高單價低銷售量的商品之召回通知。在下一篇例文中再做說明。

產品召回通知中第一會先列出問題產品的品名、有問題的出產日期，以及銷售據點。接著必須寫出問題產品會產生的影響並說明有可能會備受影響的數量。

說明了問題產品後，我們就應該說明召回產品的方式、時間、及地點。如果有問題時的連絡方式及可聯繫的時間也是絕對必要的。

A 字彙

1. **Voluntary**: (adj) 自願的，志願的
   There are many voluntary helpers at the hospital.
   醫院裡有很多自願幫忙者。

2. **Undeclared**: (adj) 未申報的
   It is illegal to bring undeclared meat into the country.
   攜帶未申報的肉品進入國家是非法的。

3. **Recall**: (v) 叫回，召回
   The famous German automaker recently recalled a lot of their vehicles.
   那個有名的德國汽車製造商近期召回許多他們的車輛。

4. **Consume**: (v) 消耗，花費；耗盡
   She consumed most of her time in taking care of her children.
   她把大部分時間都花在照顧她的小孩上。

5. **Bearing**: (n) 關係，關聯
   What you have written in the report has no bearing on the subject.
   你在報告裡所寫的與主題沒有關係。

6. **Distribute**: (v) 分發；分配
   They planned to distribute all the extra food to the local shelters.
   他們計畫將多的食物分發到地方上的收容所。

7. **Initiate**: (v) 開始；創始；開始實施

The government initiated the anti-drug promotion.

政府開始了反毒品的運動。

8. **Mistakenly**: (adv) 錯誤地；被誤解地

He mistakenly wrote his sister's address instead.

他錯寫成他姊姊的地址。

7-2 商品召回及替換通知案例二 Products recall and replacement notification

SONNOMA ANNOUNCES RECALL OF 2006 AND 2007 MODEL YEAR EAGLE HYBRID AND DELTA S VEHICLES

Detroit, Michigan., Apr. 11th, 2012 – Sonnoma Motor Sales, U.S.A., Inc., today announced that it plans to conduct a voluntary safety recall of **approximately** 43,000 Eagle Hybrid and 32,000 Delta S sold in the U.S. No other Sonnoma vehicles are involved.

The brake pedal **assembly** in the subject vehicles contains a brake pedal load sensing switch which allows the brake system to detect brake pedal force **application** by the driver. Due to a manufacturing error, there is a possibility that the switch could **improperly** cause the brake system to **activate** without driver input while driving and without activating the rear brake lights. This could result in unexpected **deceleration**.

For all involved vehicles, we will send a notification letter by first class mail advising owners make an appointment with an authorized Sonnoma dealer to have break pedal replaced with no charge. Free **leasing** car will be available under request while making an appointment. To date, Sonnoma is not aware of any accidents, injuries or **fatalities** caused by this condition.

Additional information is available to customers by calling the Sonnoma Customer Satisfaction Center at 1-800-33-1234.

產品召回和更換的通知

SONNOMA 宣布召回 2006 年和 2007 年款 EAGLE 油電混合和 DELTA S 車輛

2012 年 4 月 11 日，密歇根州，- 美國 Sonnoma 汽車銷售公司今天宣布計劃進行了於全美銷售約 43,000 台 EAGLE 油電混合車及 32,000 台 DELTA S 的自發性安全召回。無其它 Sonomma 車款包括在內。

在所提到的兩款車輛的制動踏板裝置中包含一個制動踏板載荷檢測開關，它允許在制動系統由駕駛員來檢測制動踏力的應用程序。由於製造誤差，該開關可能不當導致制動系統沒有驅動器輸入，因而後停車燈將有不被驅動的可能。這可能會導致意外的減速。

對於所有涉及的車輛，我們會發送限時專送的通知信函，通知車主與 Sonnoma 的授權經銷商預約剎車踏板的免費更換。免費租賃車可同時接受預約。迄今為止，Sonnoma 尚未被通知因這種狀況所造成的任何事故、傷害或死亡。

更多信息，請客戶撥打 1-800-33-1234 的 Sonnoma 客戶滿意中心。

達人提點 II

如例文的汽車零件召回通知是我們非常常見的。這一類的召回通知就是屬於高單價、低銷售量的產品。往往這一類型的召回是以零件更換為主，不會有“退費”的選擇。

撰寫此類文章的一開頭，往往是先列出問題產品的型號、數量以及年份。有時甚是會列舉到出貨序號。

接下來則是要說明問題點以及可能發生的狀況。由於這一類型產品召回通常都是在問題尚未發生，或發商件數非常少的情況下就發出申明，因此詳述可能發生狀況的說明可以加速顧客將問題商品更換的速度。

最後，有如一般產品召回信件一樣，產品更換的方法、時間、地點均需清楚說明。連絡窗口的聯繫方式也需在文章最後清楚列出。

A 字彙

1. **Approximately**: (adv) 大概；近乎
   There were approximately 20 cars waiting for parking.
   那裡大約有 20 台車在等車位。

2. **Assembly**: (n) 機械的裝配
   The assembly of the car parts is very complicated.
   汽車的配裝是非常複雜的。

3. **Application**: (n) 應用，適用；運用
   Surprisingly, lemon can be used for many cleaning applications.
   驚奇的是，檸檬可以被利用在很多清潔的應用上。

4. **Improperly**: (n) 不合適地 ; 不體面地
   Using any machines improperly can cause serious injury.
   不適當地使用任何機器都可能造成嚴重的傷害。

5. **Activate**: (v) 使活化
   When receiving a new credit card, you must activate it in order to use it.
   在收到新的信用卡時，你必須開卡後才可以使用。

6. **Deceleration**: (n) 減速
   While deceleration, you must watch out for the cars behind you.
   當減速時，你必須小心你後面的車輛。

7. **Leasing**: (adv) 租賃的

Leasing car is available when you bring your car for maintenance.

當你拿車去保養時，租賃車是可被利用的。

8. **Fatality**: (n) （因意外事故的）死亡；死者；死亡事故；災禍

Fortunatly, there were no fatalities in the accident.

幸運地，這個意外事故並無一死。

Unit 8 通知 Announcement

8-1 公司合併通知 Business Merger Announcement

June 15th, 2011

OBJECT: BUSINESS **MERGER** ANNOUCEMENT

Dear Sirs,

Anna's Fantasy takes great pleasure in informing you that as of June 1st, 2011, Anna's Fantasy has entered into a **definitive** merger agreement with Olaf Inc., a leading **provider** of 3D **animation**.

This merger with Olaf Inc., based in Los Angeles, CA, USA, further **broadened** and strengthened our position in the animation market. This partnership is also expected to result in better creativities, greater efficiencies, and significantly increase our market share.

As with all important business decisions, we will work closely with our partners and employees to make the **integration** process as smooth as possible. The integration of both companies will take place over the next 3 months. Of course, in the meantime, our business partners will continue to receive the same high quality services, which they have come to expect.

Olaf Inc, which will continue to operate under that name, is now a **wholly** owned **subsidiary** of Anna's Fantasy.

Sincerely,
Fiona White
C.E.O
Anna's Fantasy, Inc.

中譯

2011 年 6 月 15 日
主旨：商業合併通知

敬啟者，

Anna's Fantacy Inc. 很高興地通知您，在 2011 年 6 月 1 日，Anna’s Fantacy 已經與 3D 動畫的領先者 Olaf Inc. 完成最終的合併協議。

此次與 Olaf Inc. 的合併，我們將把總部設在美國洛杉磯，進一步擴大和加強我們在動漫市場的地位。這種夥伴關係也有望得到更好的創造力、更高的效率和顯著增加我們的市場份額。

有關所有重要的商業決定，我們將緊密的與我們的合作夥伴與員工合作，使合併過程盡可能順利。兩家公司的整合將在未來的 3 個月完成。當然，在此期間我們的業務合作夥伴將如你們所期望的繼續得到同樣優質的服務。

爾後，Olaf Inc. 將是 Anna's Fantacy 的全資子公司，但該公司將繼續以該名稱來運行。

真誠的，
Fiona White
首席執行官
Anna’s Fantacy Inc.

公司合併這一類的信件通常是公司最上層主管，例如總裁、執行長等的名義正式發出申明。

再來則是說明合併的目的、結果以及對今後的展望。如同例文中所提到的，進一步擴大和加強市場的地位，得到更好的創造力，更高的效率和增加市場份額。

在第三個段落，通常會說明合併過程所需要的時間、可能面對的問題，以及對應的方式。這是要讓員工以及客戶等相關人員安心的說明。若只是對內部的公告文，則有時會加上員工福利轉換等的資訊。

最後則是以今後的公司名稱決定作為結尾。

A 字彙

1. **Merger**: (n) 合併

   Employees are worried about the merger.

   員工很擔心公司的合併。

2. **Definitive**: (adj) 決定性的；最後的

   This kick will be the definitive play for the game.

   這一踢將會是這個球賽決定性的一擊。

3. **Provider**: (n) 供應者；提供者；供養人

   Government has become a substantial provider of the manufacturing finance.

   政府已變成製造業資金的重要提供者。

4. **Animation**: (n) 卡通片

   Animations are no longer for kids only.

   卡通片不再只是給兒童而已。

5. **Broaden**: (v) 變寬，變闊；擴大

   Working experiences often broaden one's views.

   工作經驗往往闊展一個人的見解。

6. **Integration**: (n) 整合；完成

   His profession is to help companies with their integrations.

   他的專業是幫助公司整合。

7. **Wholly**: (adv) 完全地；全部；統統

I wholly agree with his decision.

我完全同意他的決定。

8. **Subsidiary**: (n) 子公司

They have more than 20 subsidiaries worldwide.

他們在全世界有超過 20 個子公司。

8-2 退休通知
Retirement Announcement to Client

January 26, 2012
Mr. Robert Rabbit
15, Salt Lake Street
San Matel, CA

Dear Robert,

I am writing this letter to bring it to your notice that I am taking **retirement** from my position as the Quality **Assurance** Manager with Superclean Corporation, on February 15th, 2012. My job will be taken over by my new **colleague**, Jean Steward who has over 10 years of experience in **relative** firms as well as the quality assurance departments. I am fully **confident** that she will be able to keep the high standard of products for Superclean as what we have done for the past 25 years.

I will miss working for the company and with my **former** clients like you and my colleagues. And I like to thank you for your support over the past 25 years. I have enjoyed every minute working at Superclean. It was a wonderful challenge to me in many ways and **leaves nothing but** great memories. I am very glad to make great friends with you and many others.

In the future, if you have any **queries** or you wish to keep in touch with me, I will be available to you on my phone and also at mail. My number is 415-621-6732 and email is jpeterson@hotmail.com. I wish you a wonderful future and great success in your business.

Best Regards,
James Peterson

中譯

親愛的 Robert，

這封信是要通知您，我將在 2012 年 2 月 15 日由 Superclean 公司的品管經理職位退休。我的工作將交由我的新同事 Jean Steward 來擔任。她在同性質的公司及部門擁有 10 年的經驗。我充分相信，她將能夠延續 Superclean Corporation 做了 25 年的高品質產品。

我會想念於這間公司任職，還有與公司同仁和像您這樣的客戶共事的時光。我要感謝大家在過去 25 年來的支持。我很珍惜在 Superclean 工作的每分鐘。這是一個美妙的挑戰，也為我在多方面留下美好的回憶。我很開心與您和其他許多人成為好友。

今後，如果您有任何疑問，或者您希望與我保持聯繫，請以電話或電郵與我連絡。我的號碼是 415-621-6732，電子郵件是 jpeterson@hotmail.com。謹祝您在您的企業有一個美好的未來，並取得成功。

最好的問候，
James Peterson

達人提點 II

雖然說退休通知與公司合併通知乍看之下好像非常的不同，但其實寫法及文章結構大同小異。

首先要先正式說明在幾月幾號會從哪一個職位離開。接下來則是說明自己的職位將會由哪一位替代，並在信件裡說明替代者的背景及優點，以安撫客戶對於你的離開的不安。

在第三個段落中，我們通常會寫感謝的話。對曾經一起工作的夥伴，客戶等表達感謝之意。

最後則是留下自己的連絡方式，以防對於工作上有任何疑問，或者想保持連絡者沒有方式可以做聯繫。Ending 則以感謝的文字做結束。

A 字彙

1. **Retirement**: (n) 退休；退職；退役

   She plans to move to Spain after retirement .

   她計畫在退休後搬去西班牙。

2. **Assurance**: (n) 保證

   The quality assurance department plays a key role in a company.

   品保部門在公司扮演著重要的角色。

3. **Colleague**: (n) 同事，同僚，同行

   Josh doesn't get along with his colleagues.

   Josh 跟他的同事處不來。

4. **Relative**: (adj) 與……有關係的，相關的

   We are not allowed to ask questions that are not relative to the subject.

   我們不被允許問與主題不相關的問題。

5. **Confident**: (adj) 確信的；有信心的，自信的

   I am confident that I will pass the test.

   我有信心通過考試。

6. **Former**: (adj) 前任的；一度的

   As the former president of the class, she is giving a lot of good suggestions to the current president.

   前任的班長給現任班長很多好的建議。

7. **Leave nothing but**: 留下

The famous actor left nothing but good memories to all his fans.

那個有名的演員留給他的粉絲很棒的回憶。

8. **Query**: (n) 質問；詢問；疑問

If you have any queries, feel free to ask any staffs for help.

如果你有任何質問，歡迎向任何員工詢求幫助。

Unit 9 投訴處理 Complaint

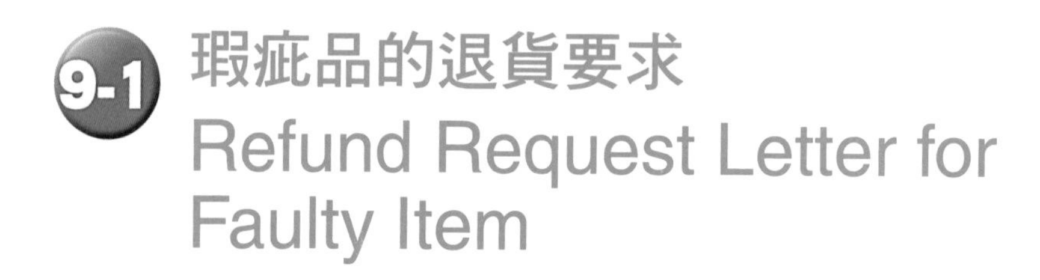

9-1 瑕疵品的退貨要求 Refund Request Letter for Faulty Item

Dear Mr. Wolfgang,

Subject: **Faulty** Wireless Vacuum Cleaner

I hope this email find you well. On July 16th 2014, I purchased the Vacuum Robot Generation 1 from your retail store in Santa Barbara Shopping Center for $39.95. The serial number is 3156732. Please find copies of my receipt and the warranty attached.

The machine was working **properly** for about a week, but I found the sucking power decreased after one week, and the engine **eventually** died in about two weeks. I checked the user manual and followed all the steps listed in the trouble-shooting guide, but nothing helped. The machine can no longer be charged and just kept making the beeping sound until the battery ran out. The machine appears to be faulty and is now unfit for its purpose. Therefore, I returned back to your store and requested for a **replacement** or a full **refund**.

The machine was **inspected** by your staff, Mary, and was found that it was **irreparable**. As it was a discontinued model and your store no longer had any in stock, Mary suggested me upgrade to the 2nd Generation Vacuum Robot for an extra $40. I considered the option but decided not to pay any extra. I requested a full refund, but Mary advised that since the Vacuum Robot Generation 1 was on sale your store does not offer refunds on sale items.

I am within my rights to request a refund for a faulty product. I would like to return the vacuum cleaner to your store for a full refund of the purchase price. I would appreciate this matter being **resolved** within the next 10 business days. If not, I am afraid I will have to forward my complaint to The Department of Consumer Affairs for further advice.

Please contact me by telephone or email at any time to discuss this letter.

Yours sincerely,
Victoria Rosenberg
(858)373-2323
vrosenberg@email.com

中譯

尊敬的 Wolfgang 先生，

主題：故障的無線吸塵器

希望您一切安好。在 2014 年 7 月 16 日，我從聖塔巴巴拉購物中心零售商以 39.9 美金購買了吸塵機器人 1 代。該序列號為 3156732。附件為我的收據影本及保固書。

該機器正常運作一個星期左右，但我發現了一個星期後，吸入功率下降，引擎最終在兩週內故障。我檢查了用戶手冊並遵循所有的故障排除指南步驟，但沒有任何幫助。本機已不能再充電，只是不停地有鳴聲，直到電池耗盡。整個機器看起來是故障的，無法達成目的。因此，我返回您的店並要求更換或全額退款。

該機是由您的員工，Mary 檢查。結果發現，這台機器無法修理。由於這個機型已經停產而您的店裡也沒有任何庫存，Mary 建議我以 40 元美金升級到第二代吸塵機器人。我考慮過，但決定不支付任何額外的費用。我要求全額退款，但 Mary 表示，由於吸塵機器人第一代是折扣商品，您的店不提供打折商品的退款。

我認為我有權利要求有缺陷的產品退款。我希望可以返回您的商店並全額退款。請在 10 個工作天內解決這件事情。如果沒有，我將不得不投訴給消費者事務署作進一步諮詢。

請通過電話或電子郵件隨時與我聯繫，以討論這封信。

此致，

Victoria Rosenberg
(858)373-2323
vrosenberg@email.com

在做投訴時，首先我們需要指出產品的購買日期、地點以及有問題的產品，甚至服務。

接下來則是說明產品的問題，為什麼需要做更換或退費。這個段落需要詳細說明問題發生的過程以及現象。越詳細的說明越能幫助對方瞭解問題的癥結。另外，如果有如例文所寫，對於服務人員也有不滿時，也應在文章的中段裡做說明。

最後則是說明希望的處理方式，包括換貨、退費等。有時在結尾時我們會加上希望處理完成的日期，例文的說法有一點威脅的成份在內，這樣可以幫助加速店家處理客訴的時間，但這也非絕對必要。

記得一定要留下自己的連絡方式，以便對方與你聯繫。

A 字彙

1. **Faulty**: (adj) 有缺點的；不完美的

   The company decided to recall all the faulty products.

   公司決定召回所有有缺陷的商品。

2. **Properly**: (adv) 恰當地；正確地

   He always speaks properly in public.

   他在公開場合總是很得體的說話。

3. **Eventually**: (adv) 最後，終於

   After one week of non-stop working, she eventually got sick.

   經過一個星期不停的工作，她終究生病了。

4. **Replacement**: (n) 代替者；代替物

   We require a replacement while the television is getting fixed.

   當電視在修理時，我們需要一個代替物。

5. **Refund**: (n) 退款；償還金額

   All on sale products do not accept refund.

   所有打折商品皆不接受退款。

6. **Inspect**: (v) 檢查；審查

   The Health Department sends staff to inspect the school every year.

   衛生部每年都會派人員來檢查學校。

7. **Irreparable**: (adj) 不能修補的 , 不能挽回的

You must pay extra attention while using this machine, because it is irreparable.

當你在用這個機器時必須格外小心，因為它是不能修補的。

8. **Resolve**: (v) 解決，解答；消除（疑惑等）

The customer service staff resolved all our questions.

客服人員解答了我們所有的疑問。

拒絕瑕疵品的退貨要求
Reject Refund Request

例 文

Dear Ms. Rosenberg,

Thank you for your letter **alerting** us to the problem you have been having with our store in Santa Barbara. I am sorry you have been **subjected** to such a **frustrating** series of events.

Unfortunately, though we regret your **dissatisfaction** with our product, we cannot offer you the refund as requested. As **stated** on the receipt, which was explained to you during your purchase, we do not accept any exchange or refund request for on sales items. Since the receipt was read and signed by you, I believe you were fully **aware of** this matter during your purchase.

With that said, as a special case, we are willing to offer you a 50% discount on the upgrade. As Mary mentioned during your visit, our standard upgrade will be $40. I will **instruct** her to give you the upgrade for $20 when you present this letter during your exchange. I will also instruct her to give you a 20% off coupon for your next purchase. I **extend** my own apologies for the inconvenience this problem has caused you.

We do appreciate your business and would be happy to see you in our store soon.

Best Regards,

Mark Wolfgang

Manager

Customer Service

中譯

尊敬的 Rosenberg 女士，

謝謝您來信告知我們您在聖巴巴拉店的問題。我為您受這樣一系列的折騰表示遺憾。

不幸的是，雖然我們瞭解您對我們產品的不滿，我們還是不能按照您的要求為您提供退款。正如在收據上所註明，並在您購買時向你解釋過，我們不接受有關折扣商品的任何交換或退款要求。由於您已閱讀及簽收了收據，我相信你在購買過程中完全瞭解這個信息。

雖然如此，我們願意以一次特殊的狀況處理，並提供您 50%的升級折扣。Mary 在您訪問期間提到，我們的標準升級是 40 美元。我將指示她在您出示這封信件時讓您以 20 美元做升級。我也將指示她給你 20%的優惠券供您下次購買時使用。對於您造成的不便，我再次表達我的歉意。

我們感謝您的光顧，並歡迎您的再度光臨。

最好的問候，

Mark Wolfgang

經理

顧客服務

在收到客戶的退換貨要求時，店家也不一定要全盤接受。有時，因為條件的問題，店家也是可以拒絕這種要求。但在拒絕的同時，我們也是需要顧慮到將來生意的可能性。因此，在信中我們大約會做這樣的回覆。

第一，先告知對方已收到投訴信函，並予以處理及討論。並對於所造成的不便表示遺憾。

接下來，則是說明不能接受退換貨的原因。這個部份也是與投訴信函相同，說明越詳細越能說服對方。如例文中所提到的，「正如在收據上所註明，並在您購買時向你解釋過，我們不接受有關折扣商品的任何交換或退款要求。由於您已閱讀及簽收了收據，我相信你在購買過程中完全瞭解這個信息。」，這樣的說明足以讓對方瞭解退貨的要求是不可能的。

最後，商家的部份當然希望留位這個客戶，因此我們會以合作的態度結尾。有如例文中提及的一些特殊折扣，這樣可以幫助留下這個客戶。

A 字彙

1. **Alert**: (v) 通知

   The weather channel just alerted about the upcoming storm.

   氣象台剛剛通知了即將來臨的暴風雨。

2. **Subject**: (v) 使蒙受；使遭遇

   He is upset that his family had subjected to such a tragedy.

   他很難過他的家人遭遇到如此的悲劇。

3. **Frustrating**: (adj) 惱怒的；不滿意的

   The store manager apologized for the frustrating situation.

   店長對於惱怒的狀況道歉。

4. **Dissatisfaction**: (n) 不滿；不平

   All staffs are required to review customers' dissatisfactions everyday.

   所有人員都被要求每天要檢閱客人的不滿。

5. **State**: (v) 陳述；聲明；說明

   As stated on the menu, all coffee is freshly grinded every morning.

   如同菜單上所陳述的，所有的咖啡都是當天現磨的。

6. **Aware of**: 充分了解或意識到

   We cannot let you join the project, unless you are well aware of the risks.

   除非你清楚知道其中的風險，不然我們不能讓你參加這個計畫。

7. **Instruct**: (v) 指示，命令，吩咐

My mother instructed me to visit you today.

我的母親吩咐我今天過來拜訪你。

8. **Extend**: (v) 延長，延伸；擴大，擴展

I am asking to extend my vacation from 3 days to 7 days.

我請求將我的假期由三天延長到七天。

Unit 10 報告及記錄 Reports/ Minutes

10-1 會議紀錄 Meeting minutes

例 文

Organic Daily, Corp.

Board Meeting Minutes: June 4th, 2014

9:00 am

San Francisco, CA

Board Members:

Present: Grace Hung, Neilson Wolfe, Ryan Pan, Edward Chen, Kanae Kumamoto

Others Present:

Exec. Director: Sarah Chen

Proceedings:

● Minutes of previous meeting was amended and approved

● Chief Executive's Report:

- New facility has been found and will sign the contract by the end of the month. Currently, facility will be tore down by

September 31st notified by the county. Therefore, no delay of moving can be accepted.

- Sales group **exhibited** the OCC Trade Show in Atlantic City, NJ last month. Organic Dry Fruit Chips were **released** during the show. **Feedbacks** from exhibition **attendees** will be reviewed within this month and the result will be reported in the next board meeting.

- Asian Pacific sales head reported a continuous increase in **revenue** for the past quarter and suggested expanding 10 branches in 3 cities- Seoul, Tokyo, and Osaka. Board members agreed. Requests detailed business plan to be presented in the board meeting next month.

●Finance Committee report provided by Chair, Neilson Wolfe:

- Wolfe reviewed highlights, trends and issues from the balance sheet, income statement and cash flow statement.

●Other business:

- Pan noted that he was working with tech department to update the website. Online shopping site will be upgraded to a more customer friendly design within 3 months. Demo page will be presented to board members in the next meeting.

- Chen announced that new contract with hiring service company Job.com will be **renewed** by the end of the month.

●Minutes **submitted** by secretary, Barbara Smith.

中譯

Organic Daily, Corp.

董事會會議記錄：2014 年 6 月 4 日

上午 9:00

舊金山，加州

董事會成員：

出席者：Grace Hung、Neilson Wolfe、Ryan Pan、Edward Chen、Kanae Kumamoto

其他出席者：

執行長：Sarah Chen

議程：

●上次會議記錄已修訂和批准。

●行政長官的報告：

- 新的辦公室已找到，月底前將簽署合同。政府通知，目前的辦公室將在 9 月 31 日拆除，因此，辦公室的遷移不能被延誤。

- 銷售集團上個月在紐澤西洲大西洋城的ＯＣＣ貿易展展出。有機乾果片是在展會期間發布。與會者的反應將在本月內審查，結果將在下次董事會會議上匯報。

- 亞太地區的銷售負責人報告了過去一個季度的持續收入增長，並建議在首爾、東京、大阪三個城市增開 10 家分店。董事會成員表示同意並要求在下個月董事會會議時提出詳細的商業計劃。

●財務委員會報告，財務長 Neilson Wolfe 提供：

- Wolfe 審查資產負債表、利潤表和現金流量表的亮點、趨勢和問題。

●其他業務：

- Pan 提出他與科技部門合作更新網站。線上購物網站將在 3 個月內升級，設計將更便於消費者使用。展示網頁將在下次董事會提出。

- Chen 提出將在本月底與招聘服務公司 Job.com 續約。

●報告提交者：秘書 Barbara Smith。

會議記錄的目的是將會議記錄下來，讓內容可以被出席者確認。另外，未參加會議的人員也可以由會議記錄中瞭解會議的狀況及內容。

一般在做會議記錄時首先必須列出開會的時間及地點、出席者，以及會議背景。

出席者的全名及部門也須清楚紀錄，以瞭解責任歸屬。

會議記錄應該使用"摘要式"紀錄。不要使用太長的句子，紀錄應簡潔有力，長度應該控制在一頁以內。

開會的目的就是為了進行討論及取得結論，因此進行會議記錄必須注意下列要點：

1. 討論事項及主題　2. 決議事項及各項決議的完成時間點

3. 各單位負責事項及內容　4. 下次會議時間、主席及議題

會議後應發郵件請大家確認會議記錄內容。

A 字彙

1. **Proceedings**: (n) 會議記錄

   It is my turn to take the proceedings today.

   今天輪到我寫會議記錄。

2. **Exhibite**: (v) 舉辦展示會；展出產品

   This will be the first time for our company to exhibit in this show.

   這將是我們公司第一次在這個秀裡展出。

3. **Release**: (v) 發行；發表

   Many people are lining up at the theater now because a new movie is to be released tonight.

   很多人現在都在電影院排隊，因為今晚要發行一部新電影。

4. **Feedback**: (n) 回覆 , 反應

   It is always important to get feedbacks from customer in order to improve our services.

   接收客戶的反應是重要的，這樣才能改進我們的服務。

5. **Attendee**: (n) 出席者；在場者

   All the attendees today will receive a special gift.

   今天所有的出席者都將收到一個禮物。

6. **Revenue**: (n) 各項收入，總收入 [P]

   An increase in revenue doesn’t mean an increase in profit.

   總收入的增加不代表利益的增加。

7. **Renew**: (v) 使更新；使復原；使恢復

The company website will be renewed after the logo change.

公司的網頁將在商標更換後更新。

8. **Submit**: (v) 提交，呈遞

I just submitted the report.

我剛剛提交報告。

10-2 月報 Monthly Sales Report

Monthly Sales Report		
Covered Month: Aug, 2008	***Report By: Leslie Hsu***	***Report Date: 09/12/08***
Visited Customer List: ABC - Sunnyvale, CA StarTech - Santa Clara, CA	Estelic - Albuquerque, NM	
High-Lights:		
ABC: - Add on order for model -584. - Due to the market demand increase, ABC requests a 2 weeks stock in hand for all time. In return, **forecast** will be updated weekly instead of monthly. StarTech: - Model #33445 has been **qualified**. - Sample run will start within one week. If sample run passes the quality test, 1st production run will start in 2 months. - Requested updated **quotation** in annual volume. - Currently 3 **vendors** are **bidding** on the same project. 1 vendor will be chosen as a **primary** supplier which supplies 70% of the demand, and one more vendor will be chosen as 2nd supplier which supports 30% of the demand. The 3rd supplier will only listed as back up. - Vendor qualification judgment: Quality 40%, Unit Price 30%, Delivery 30% Estelic: - Requests cross-reference from competitor Alpha, part 9986. - Exact same spec cannot be provided by us due to copyright, but similar specification can be manufactured. - Requested spec will be sent to design engineer department for review and respond back to customer within one week.		
Competitor Update: N/A		

Request to Factory:			Factory Respond	
New Item		Request Date		Respond Date
1) Lead time update for model -584 need to be shortened from 4 weeks to 2 weeks.		8/12/08	1) Confirmed	9/4/08
2) StarTech production cost review.		8/22/08	2) Replied.	8/26/08
3) Estelic cross-reference review.		8/28/08	3) **Rejected**. Requested spec is out of current **capability**. Requests sales to re-negotiate the specification.	9/10/08
Previous action item update:				
1) Beta Model -5567 production problem		7/1/08	1) Model -5567 is not longer in production and requests customer to change to the new model.	8/10/08

Actual Sales vs. Sales Forecast

	Jul 08	Aug 08	Sep 08	Oct 08	Nov 08	Dec 08	Jan 09	Feb 09
Sales Target	\$250,000	\$250,000	\$250,000	\$250,000	\$250,000	\$250,000	\$250,000	\$250,000
Sales Forecast	\$206,000	\$204,000	\$220,000	\$250,000	\$260,000	\$200,000	\$200,000	\$200,000
Actual Sales	\$264,583	\$295,790						

中譯

月報		
涵蓋月份 : 2008 年 8 月	提報者 : Leslie Hsu	提報日 : 2008 年 9 月 10 日
已拜訪客戶名單 : ABC - Sunnyvale, CA StarTech - Santa Clara, CA Estelic - Albuquerque, NM		

提要：

ABC:

- 追加 型號 -584 的訂單 .
- 由於市場需求增加, ABC 要求本公司隨時準備兩星期的庫存。對方的用量預估也會由月更新改為週更新作為回報。

StarTech:

- 型號 #33445 已通過認可。
- 下週將進行打樣。若打樣商品通過測試，兩個月內將開始第一次量產。
- 要求更新年度報價。
- 目前有三個供應商在競標相同的案子。第一供應商將成為主要供應商並獲得 70% 的份額。第二供應商將得到剩下的 30% 份額。第三供應商將成為備選供應商。
- 供應商選擇將會以品質 40%、單價 30%、送貨 30% 為基準。

Estelic:

- 要求提供與競爭對手 Alpha 品項 9966 相同規格的產品。
- 由於版權的關係，無法提供完全相同規格的產品，但可製造類似規格的產品。
- 客戶要求的規格將被送去給設計部門做檢閱並在一週內回付給客戶。

競爭對手資訊 :

無

給工廠的要求 :		工廠回覆	
新品項	要求日期		回覆日期
1) 型號 -584 的交期需由 4 週縮短為 2 週	8/12/08	1) 確認	9/4/08
2) StarTech 生產費用評估	8/22/08	2) 已回覆	8/26/08
3) Estelic 要求規格檢討	8/28/08	3) 拒絕。客戶所要求的規格已超出公司的技術能力。 請業務部門再次與客人交涉規格需求。	9/10/08
前次行動項目追蹤 :			
1) Beta 型號 -5567 生產問題	7/1/08	1) 型號 -5567 已停產並請客戶轉換使用新的型號。	8/10/08

實際銷售值及預估銷售值的比較

	2008 年 1 月	2008 年 2 月	2008 年 3 月	2008 年 4 月	2008 年 5 月	2008 年 6 月	2008 年 7 月	2008 年 8 月
銷售目標	$250,000	$250,000	$250,000	$250,000	$250,000	$250,000	$250,000	$250,000
銷售預估	$206,000	$204,000	$220,000	$250,000	$260,000	$200,000	$200,000	$200,000
實際銷售值	$264,583	$295,790						

達人提點 I

月報的目的是讓與你工作的人員瞭解這個月內你的工作內容以及是否有需要他方支援的一大工具。與會議記錄相類似的地方是在做這類行的報告時，我們都傾向使用條例式的寫法，這樣的書寫方式可以減少贅字，並簡短閱讀時間。

內容的排列部份則與一般書信類似。

第一部份：列出報告涵蓋月份、提交人員，及提交日期。

第二部份：條例出這個月的重點提要，並以簡單的文字說明內容。

第三部份：說明需要其它部門支援的項目，並說明前月要求支援的進展。

最後則是列出相關數字的說明。例文是以業務的角度撰寫的月報，因此數字的報告則是以銷售相關為主。若是其它部門，則可做不同的調整。

A 字彙

1. **Forecast**: (n) 預報，預測，預料
   The weather forecast shows it is going to rain tomorrow.
   天氣預報顯示明天將下雨。

2. **Qualified**: (adj) 合格的；勝任的
   She is barely qualified for the contest.
   她勉強合格能參加這比賽。

3. **Quotation**: (n) 報價單
   The quotation was submitted to the customer yesterday.
   報價單昨天已交給客戶。

4. **Vendor**: (n) 賣主；供應商
   It is an honor to be qualified as their vendor.
   成為他們的賣主是一種榮耀。

5. **Bid**: (v) 喊價，出價；投標
   The company lost at the bidding on the navigation system.
   公司在投標導航系統裡輸了。

6. **Primary**: (adj) 首要的，主要的
   Our primary goal is to finish the project on time.
   我們首要的目標是在時間內完成計畫。

7. **Reject**: (v) 駁回；否決
   His business proposal was rejected due to lack of information.
   由於缺乏資訊，他的商業提案被駁回了。

8. **Capability**: (n) 能力，才能

She was hired because of her language capability.

她是因為她的語文能力被雇用。

Unit 11 休假請求 Vacation Request

11-1 產假請求 Maternity Leave Request Letter

Dear Mr. Koyama,

I am writing this letter to inform you that I wish to take **maternity** leave starting from May 1, 2015. As you are **aware** I am 8 months pregnant and the baby is due around the end of May 2015. I have attached the **relevant** medical application and other details. I would like to **avail** the entire maternity leave and benefits. Please guide me through the process of applying for the **benefits**.

The leave I would like to apply from May 1, 2015 to July 31, 2015. I also want to add up my annual leave along with the maternity leave so that I can be with the baby after the delivery. All in together I will be taking a 3 months leave and will join **immediately** after that. But in case of any complication I will inform you about the **extension** of the leave. During my leave, Ms. Angela Megano will cover my job, and I can be reached by cell phone and email at anytime. Please feel free to contact me if any assistance is needed.

I hope you will **consider** my situation and allow me to take the annual leave along with the maternity leave. Please confirm me the dates of leave and also the date of joining back.

Waiting for the reply and thanks for your time and consideration.

Sincerely,
Amber Channing
Employee No: 21211
Dept: Sales & Marketing

中譯

尊敬的 Koyama 先生，

這封信是希望能知會您我希望能從 5 月 1 日起休產假。如同您所知的，本人目前懷孕 8 個月，且預產期為 5 月底。我已附加了相關的醫療申請書和其他細節。我想使用整個產假和福利。請指導我通過福利申請。

我想申請的假期為 2015 年 5 月 1 日至 7 月 31 日。同時，我也希望可以加入我的年假申請連休，這樣我可以在分娩後繼續陪伴我的寶貝。所有假期加在一起我將離開 3 個月，隨後會立即返回崗位。但是，如果有任何突發狀況，我將會通知您有關休假的延長。在我的離開的期間，Angela Megano 女士將代替我的工作，而我則可通過手機和電子郵件被隨時連絡到。如果需要任何協助，請隨時與我聯繫。

我希望你能考慮我的情況，並讓我加入我的年假，讓我能休產假時一同休年假。請幫忙確認我可以休假的日期，及返回工作的日期。

等待您的答覆，並感謝您的時間和體貼。

真誠的，

Amber Channing

員工號碼 : 21211

部門 : 銷售與市場營銷

達人提點 I

在工作中，往往會有需要提出休假請求的時候。撰寫休假要求時所需注意的提點如下：

首先在信件中要提到的是休假的開始日期以及要求休假的原因。若是醫療相關的休假，則建議提出相關的資料做為備份。若為私人休假則不需要。

接下來在內文中，應說明希望休假的始末日期及天數。並告知代理人是哪位，以及為何他可以勝任代理人的職務。同時也必須說明是否在休假期間可被連絡。如果可以被連絡，則應說明連絡方式及連絡時間。這些資訊都可以讓主管更放心地讓你 take vacation，也會讓你收假後回到辦公室的氣氛更和諧。

字彙

1. **Maternity**: (adj) 孕婦的；產婦的
   Maternity products are a growing market.
   孕婦商品是一個成長中的市場。

2. **Aware**: (adj) 知道的，察覺的
   She was not aware that the movie is not suitable for kids.
   她不知道這部電影不適合小孩。

3. **Relevant**: (adj) 有關的；切題的；恰當的
   His behavior is not relevant to whether he grew up in a good family.
   他的行為與他是不是生長在好的家庭不相關。

4. **Avail**: (v) 有用於；有益於；有助於
   She availed the chance to get the promotion.
   她利用這個機會以得到升遷。

5. **Benefit**: (n) 津貼，救濟金
   This company gives an extra day off as a special benefit.
   這間公司給出多一天休假當做特殊津貼。

6. **Immediately**: (adv) 立即，即刻，馬上
   Students are required to leave the building immediately.
   學生被要求要馬上離開建築物。

7. **Extension**: (n) 延長；延期；緩期

The vendor asked for the extension of the deadline to the end of the month.

供應商要求把最後期限延至月底。

8. **Consider**: (v) 考慮，細想

He is considering taking his degree in Japan.

他正在考慮去日本拿他的學位。

11-2 留職停薪休假請求
Leave Without Pay Request

Mr. John Lee,
Personnel Manager,
AAA Real Estate
June 15, 2011

Dear Mr. Lee

I am writing to request an **unpaid** leave starting from Monday, July 1 to July 7. I have just returned from my annual leave and do not have any more off days left for this year. However, I **urgently** need to go out of the country to **extend** my working VISA. I have discussed with the **legal** department, and unfortunately, it is not possible to extend the VISA while I am in the country. All working VISA **applications** need to be done overseas. Therefore, if I do not take care of this matter now, I will no longer be able to work legally in this country.

I **recommend** that Debra Hopkins to take over my role until I return. Debra and I have very similar job responsibilities, and she is **familiar** with my clients. She is also the person who covered my role during my annual leave. I think she will be the most **suitable** person to help me with my job. Also, I will not be completely out of contact and you can reach me by phone at any time during my leave period.

I appreciate your cooperation in this matter.

Sincerely,
Medlin Chen

中譯

John Lee 先生，
人事經理，
AAA 房地產
2011 年 6 月 15 日

親愛的李先生，

這封信是本人請求可以從 7 月 1 日星期一起，至 7 月 7 日止申請無薪假。我剛才從我的年假回來了，沒有留下更多的休假。不過，我迫切需要出國以延長我的工作簽證。我曾與法律部門討論，不幸的是，我無法在本國內延長我的工作簽證。所有的工作簽證申請皆需要在海外完成。所以，如果我現在不處理這件事情，我將無法再繼續於這個國家合法工作。

我建議 Debra Hopkins 接管我的工作，直到我回來。Debra 和我有著非常相似的職責，她也熟悉我的客戶。同時，她也是在我年假時代替我的角色的人。我想她會是最合適幫助我工作的人選。另外，我也不會完全脫離聯繫，您可以在我休假期間的任何時候通過電話連絡我。

我感謝您在這個問題上的合作。

真誠的，
Medlin Chen

除了一般的假期申請外，我們有時也會因為特殊或臨時的狀況，而需要臨時請假或動用到無薪假。這時，撰寫無薪假申請時又該注意什麼呢？說實在的內容大同小異，但這時因為是特殊的要求，所以“原因”的闡述就非常重要。

同樣的，在開頭我們必須寫出希望休假的日期及天數。例文中是以起始日期與結束日期作為闡述。接著則須說明為何迫切的需要請無薪假，這個部份的說明必須非常明確。在說明後，當然需要列出自己的代理人，以及自己可以被連絡到的方式，這個部份與一般休假申請是相同的。

在最後，由於是臨時的要求，所以簡單的感謝句將是個很好的收尾方法。

A 字彙

1. **Unpaid**: (adj) 無報酬的

We are his unpaid employees this weekend.

這個週末我們是他不支薪的員工。

2. **Urgently**: (adv) 緊急地，急迫地

She urgently asked for water.

她急迫地要了開水。

3. **Extend**: (v) 延長，延伸；擴大，擴展

The doctor instructed him to extend his sick leave.

醫生指示他延長他的病假。

4. **Legal**: (adj) 法律上的，有關法律的

He has no choice but to take a legal action.

他沒有其它選擇，只能提出訴訟。

5. **Application**: (n) 申請，請求；申請書

The deadline to turn in the college application is today.

今天是繳交大學申請書的期限。

6. **Recommend**: (v) 建議，勸告

I recommend you go to that new restaurant at the corner.

我建議你去轉角的那家新餐廳。

7. **Familiar**: (adj) 熟悉的，通曉的

My mother is familiar with the job.

我母親很熟悉這個工作。

8. **Suitable**: (adj) 適當的；合適的；適宜的 [（+to/for）]

No one is more suitable than him to the job.

沒有人比他更適合這份工作。

Unit 12 商務活動 Commercial Activities

12-1 邀請觀展 Invitation to the Trade Exhibition

Invitation - Visit our Booth at IMS 2015

Dear Mr. Gonzales,

RE: International Microwave **Symposium** – 17-22 May 2015, Phoenix

On **behalf** of Electronic Design Studio, please allow me to **extend** an invitation to you to visit our booth during the IMS Trade Show as mentioned above. At our booth, we will have displays of our multi channel high speed modulator and will be **conducting demonstrations**. Our representatives will be there to answer any questions you may have about our services.

Electronic Design Studio has been offering our module design service to global well-known customers such as Intense

Microwave / Moon Microsystem / Southwest Instruments/ etc. for over 20 years. Our clients **consistently** receive first rate service from us and we stand behind our services with our 10 year-warranty. No customer is too big or too small for us to **accommodate**.

We invite you to take advantage of the displays and demonstrations that will be presented at our booth. Please be sure to stop by. Enclosed, we are pleased to provide you with details of the IMS 2015 including **venue**, dates and the specific location of our booth for your convenience.

We look forward to seeing you there!

Yours truly,
Emily Wood
GM
Electronic Design Studio

邀請函 – 請參觀我們在 2015 年ＩＭＳ的展位

尊敬的 Gonzales 先生，

RE：國際微波研討會 - 2015 年 5 月 17 日至 22 日，鳳凰城

謹代表電子設計工作室，請允許我邀請您在上述 IMS 展會期間來參觀我們的展位。在我們的展台，我們將有我們的多通道高速調製器的展示並將進行現場演示。我們的代表將在那裡回答您對我們的服務上的任何疑問。

電子設計工作室一直為全球熟知的客戶從事模塊的設計服務，如 Intense Microwave / Moon Microsystem / Southwest Instruments 等，為期超過 20 年。我們的客戶不斷地接受我們一流的服務，我們替我們的服務背書並提供 10 年質保。對我們來說，我們的服務沒有客戶太大或太小的分別。

我們敬邀您來參加，並藉此參觀我們將在我們的展台所呈現的顯示器和演示。請務必前來。我們很高興為您提供 IMS 2015 年的詳細信息隨函附上，包括地點、日期和我們的展台位置，以方便您前來。

我們期待著您的光臨！

敬上，

Emily Wood

總經理

電子設計工作室

達人提點 I

這是寄給客戶邀請前往展覽的邀請函。我們必須在邀請函中引起客戶的興趣，才能促使客戶前往。最好的方式就是告知展場上會有什麼特殊的展品。

首先在邀請函的開頭必須先列出展覽的名稱、時間、地點。這是非常重要的。

接著，邀請函中則須說明什麼樣的產品會在這次的展覽中展出，特殊的點在哪裡。記住，這個部份必須要有辦法提起客戶的興趣，客戶才有可能會前來。在邀請函中也建議簡單說明公司的介紹，並說明公司的強項。

最後則別忘了再次歡迎客戶前來參觀。

A 字彙

1. **Symposium**: (n) 討論會；座談會

   We are assigned to be the staff at the symposium.

   我們被指定為研討會的工作人員。

2. **Behalf**: (n) 代表；利益

   This flower is sent on his behalf.

   這個花是代表他送的。

3. **Extend**: (v) 致；給予，提供

   I would like to extend a warm welcome to our visitors.

   我想對我們的來訪者表示熱烈的歡迎。

4. **Conduct**: (v) 實施；處理；經營，管理

   He was specially hired to conduct her tax affairs.

   他是特別被請來處理她的稅務。

5. **Demonstration**: (n) 實地示範，實物宣傳

   It is always easier for students to understand a theory when the teacher teaches by demonstration.

   當老師示範教學時，學生總是比較容易理解。

6. **Consistently**: (adv) 一貫地；始終如一地；不斷地

   He consistently donates money to the hospital.

   他不斷地捐錢給醫院。

7. **Accommodate**: (v) 照顧到，考慮到

While accommodating seniors, you need to pay extra attention.

當照顧老人時，你必須特別注意。

8. **Venue**: (n) 發生地；集合地

HR will announce the venue at lunch.

人事部將會在午餐時間宣布集合地。

12-2 展覽介紹 Introduction of Exhibits

Join us at the World's Number One fine watch event in Geneva

Geneva, the world's **premier** trading **hub** for fine watch is a city that gathers all world top designers and **skilled** workers where there are no better places in the world you can find better quality or designs for watches. It is also the ideal springboard from which traders worldwide can seek for the most rare **vintage** collections.

The Fine Watch Fair in October, **organized** by V&F, continues to attract key players in the world's watch industry, a **hallmark** of a truly successful fair. In 2012, the October Fair hosted a select group of around 300 exhibitors from 25 countries and regions. They **occupied** more than 25,000 square metres of exhibition space. The fair welcomed over 35,000 visitors from around the world. The attendance figure attests to the fair's position as a vital fine watch marketplace that every serious fine watch collector and **connoisseur** cannot afford to miss.

中譯

請加入我們在日內瓦的世界頭號精品手錶展

日內瓦，這個世界上首屈一指精品腕錶的貿易樞紐是集所有世界頂尖的設計師和技工的地方，世界上沒有比這更好的地方可以找到質量或設計更好的手錶。這也是從世界各地，尋找稀有古董收藏品交易者的理想跳板。

V & F 組織在十月所舉辦的精品鐘錶展持續吸引世界鐘錶業的關鍵人物，這是一個真正成功博覽會的一大特點。 2012 年 10 月份，此博會特選了來自各國家及地區，約 300 家參展商出展。他們佔據了超過 25,000 平方米的展覽空間。此活動歡迎了來自世界各地超過 35,000 人次的訪客。出席數字證明了展會的地位，是每一個認真的鐘錶收藏家和鑑賞家不能錯過的重要精品手錶市場。

達人提點 I

這篇展覽邀請與前篇展覽邀請不同的地方在於：前一篇的展覽邀請是寄給個別客戶的邀請函，而這篇例文的展覽邀請則是公告式的寫法，為了吸引更多的人去參加這個活動。

在撰寫內容方面則是以整個展覽的內容為主。我們必須道出這個展覽的特點，有什麼樣的人會特別來參觀這個展，並說明有參展的好處，有如例文裡所說的"這個展覽是可以尋求最難得的古董收藏品的理想跳板"。

最後，則可說明過去展覽的成功案例，更增加一般人前來參觀的興趣。

字彙

1. **Premier**: (adj) 首位的；首要的
   This is the premier airport of the United Sates.
   這是美國的第一大機場。

2. **Hub**: (n) 中心；中央
   My company uses the factory in Thailand as a shipping hub.
   我們公司利用泰國的工廠為運輸中心。

3. **Skilled**: (adj) 熟練的；有技能的
   A skilled technician is not easy to find.
   要找一個熟練的技師是不容易的。

4. **Vintage**: (adj) 古色古香的；古老的

Vintage cars are cool to drive but hard to maintain.

老爺車開起來很酷但很難保養。

5. **Organize**: (v) 組織；安排

The student union is starting to organize the prom.

學生協會開始安排畢業舞會。

6. **Hallmark**: (n) 標誌，特徵

His return is a hallmark of success.

他的歸來是成功的標誌。

7. **Occupied**: (v) 已佔用的；在使用的；無空閒的

Her time is all occupied with her kids' activities.

她的時間都被她小孩的活動所佔用。

8. **Connoisseur**: (n) 鑑賞家，行家

It takes years of experience to become a connoisseur.

成為一位鑑賞家需要多年的經驗。

Unit 13 離職 Resignation

13-1 離職函 Resignation Letter

Amy Green, Sales Director
ForeverGreen Country Club
9033 Albermarie Rd,
Charlotte, NC28227

Dear Ms. Green,

Please accept this written **statement** as an advance notice of my **resignation** from ForeverGreen Country Club. I **intend** to work through the end of May with my last working day of Friday, May 30th. Please know that I am **committed** to remaining a part of the company and continuing to do whatever it takes to assist the customers till the end.

The decision to leave ForeverGreen Country Club after almost 5 years of service has not come easy. I love the duties of this job and the customers I have gotten to know over the years. As you know, I have been planning to take working holidays in Australia for many months now and the time has nearly arrived to **depart**. As per our conversation last month, I do understand why my request to keep the job without pay for one year was **denied**. But for me, this is an opportunity of a lifetime – one that I cannot afford to pass up for many reasons and one that, **consistent** with my life values and priorities must be done no matter the cost.

I sincerely thank you for the opportunities, experience, and knowledge I have gained during my career at ForeverGreen Country Club. I have so enjoyed working with many of the people who work directly under your supervision. If there is anything I can do to provide a smoother **transition** during my remaining weeks, please feel free to let me know.

Sincerely,
Sherry Wang

中譯

Amy Green 銷售總監
ForeverGreen 鄉村俱樂部
9033 Albermarie Rd,
Charlotte, NC28227

尊敬的 Green 女士，

請接受這份我從 ForeverGreen 鄉村俱樂部辭職的書面提前通知。我打算工作到五月的最後一個星期五，5 月 30 日將會是我的最後一個工作日。請瞭解我將會持續做為公司的一份子，並繼續盡一切力量協助客戶直到結束。

決定要離開服務將近五年的 ForeverGreen 鄉村俱樂部並不容易。我熱愛這份工作的職責，也熱愛認識多年的客戶。如您所知，這幾個月以來，我一直為我的澳洲打工假期計劃，現在也已快到了離開的時間。按照我們的上個月的談話，我明白我留職停薪的請求遭拒絕的原因。但對我來說，這是一個千載難逢的機會 - 我不能放棄的原因有很多，這是其中一個其另一個是不管付出任何代價，我都要秉持一貫的生活價值觀和優先順序。

我真誠地感謝您讓我在 ForeverGreen 鄉村俱樂部的職場裡所獲得的機會、經驗和知識。我是如此地享受與在您監督下工作的人一起做事。在剩下的幾個星期內，如果有什麼我可以做到以完成平順的過渡期，請隨時讓我知道。

真誠的，
Sherry Wang

撰寫離職信往往都讓人很困擾，不知如何下手。英文的離職信相對來說有一般制式化的寫法。“Please accept this written statement as an advance notice of my resignation. ”幾乎是固定的開頭文。想當然耳，接下來則是說明離職日期。一般來說，在歐美系國家會給與大約兩週以上的時間讓公司作準備。

接下來的部份雖非必要，但出自於商業禮貌，離職者還是會加以說明離職的原因。離職原因有很多種，並不侷限於特殊理由。

最後則是感謝在這份職場所學到的經驗，以及表達會在離職前做好自己份內的工作，並幫助交接。

A 字彙

1. **Statement**: (n) 正式的聲明

   The president will soon make his public statement about the election.

   總統很快會就選舉發表公開聲明。

2. **Resignation**: (n) 辭職書，辭呈

   I just turned in my resignation.

   我剛剛遞出辭呈。

3. **Intend**: (v) 想要；打算

   He intends to marry his girlfriend this year.

   他打算今年和他女友結婚。

4. **Commit**: (v) 使承擔義務；使作出保證；使表態

   The solders commit themselves to the country.

   軍人對國家做出保證。

5. **Depart**: (v) 起程，出發；離開，離去

   They departed yesterday to Paris.

   他們昨天出發去巴黎。

6. **Deny**: (v) 拒絕給予；拒絕的要求

   Their request to go camping was denied.

   他們去露營的要求被拒絕了。

7. **Consistent**: (adj) 始終如一的，前後一致的

His behavior remained consistent with his speaking.

他的言行如一。

8. **Transition**: (n) 過渡；過渡時期

Eveyone needs to pay extra attention to their work during the transition of merger.

在公司合併的過渡時期，每個人都必須格外注意自己的工作。

13-2 雇主接受（離職）函 Acceptance Letter

例 文

Date: May 10th

Dear Sherry,

We accept your resignation from the company **effective** on May 30th, with regret. We appreciate your **generous** offer to continue to work until the end of the month.

Please turn over your keys and other company **property** on May 30th before leaving. We appreciate your **willingness** to answer any questions we may have in the near future by phone or email. When you have turn over all company property, please have your **supervisor** sign the acknowledgment below.

We will instruct **payroll** to make May 30th your last day of work with regard to all pay and benefit issues. If you wish to take **advantage** of our legal requirements to keep you on our medical plan for such period of time, please **indicate** that below so we can make the proper arrangements for our mutual protection.

We also appreciate you signing the Employee Final Release so that we have all of your paperwork in order.

Our thanks for your good work,

With best regards,

\---

Authorized Employee of Company Acknowledgement that all Property Returned by Authorized Supervisor

I wish to exert my legal right to continue receiving Medical Benefits, and pay for them accordingly, until I give the company further notice that I no longer want to take advantage of this legal right of mine.

\--------------------------

Employee

中譯

日期：5 月 10 日

親愛的 Sherry，

我們帶著遺憾接受您的辭呈從 5 月 30 日起生效。我們感謝您的慷慨，願意繼續工作到月底。

在 5 月 30 日當天，請繳交您的鑰匙和其他公司的財產後才能離開。感謝您願意在不久的將來能通過電話或電子郵件替我們解答我們任何的疑問。當你交出所有公司財產後，煩請您請您的主管簽署以下的確認單。

關於工資及福利，我們將通知薪資單位將您的工資及福利計算到 5 月 30 日。如果你想利用我們的法律規定，在這段期間內繼續接受我們的醫療計劃，請在下列說明，以便我們可以相互保護並妥善安排。

我們也感謝您簽訂員工最終放行單，這樣，我們所有的文件都就緒了。

我們感謝您好的工作表現，

最誠摯的問候，

公司員工授權　　　　　　確認所有公司財產繳回的主管

我希望繼續接受醫療福利，並支付相關費用，直到我進一步通知公司我不再想接受這種法定利益。

僱員

達人提點 II

當收到員工的離職信函時，雇主方則需發出一張接受離職的文件。這份文件裡說明了瞭解員工即將在某日離職並說明雇主方需與即將離職員工確認的事項。

開頭，應該說明離職的生效日期。你會發現，在英語的商用文件裡，幾乎所有的文件都會以日期及內容大綱確認為開端。這是避免誤會的一大準則，與亞洲國家的撰文方式有些不同。

本文部份則說明了離職前公司需要員工繳回的文件及物品、需繳回的時間，且需與哪一位主管確認。

在最後則是說明薪資及福利的計算，並告知員工在離職時須簽下所需的文件。

A 字彙

1. **Effective**: (adj) 生效的，起作用的

   The credit card becomes effective as soon as you make the phone call.

   信用卡在你打過電話後便會生效。

2. **Generous**: (adj) 慷慨的，大方的

   She is rich but not generous in giving help.

   她非常有錢但不樂於助人。

3. **Property**: (n) 財產，資產；所有物

   He sold all his properties before the divorce.

   他在離婚前賣掉了他所有的資產。

4. **Willingness**: (n) 自願；樂意

   I appreciate your willingness to help the elderly.

   我很開心你自願幫助老人。

5. **Supervisor**: (n) 監督人；管理人；指導者

   As of this month, she was promoted as the supervisor of the department.

   這個月起，她被升遷為部門的管理人。

6. **Payroll**: (n) 薪水帳冊；發薪名單

   If you have any questions in regard to payroll, you should contact your supervisor.

   如果你對薪水帳冊有任何問題，請連絡你的主管。

7. **Advantage**: (n) 利益，好處

It is not nice to take advantage of the children.

占小孩便宜是不好的。

8. **Indicate**: (v) 指示；指出

Her letter indicates that she will not come to work anymore.

她的信件指出她將不會再來上班。

Unit 14 付款協商 Payment Negotiation

要求延遲付款 Request a delay on payment

August 20th, 2013
Wild and Fresh Seafood Corp.
No.212 Da-An Rd,
Da-An Disctrict, Taipei 106

Gentlemen,

Hotel Rosenberg appreciates your partnership for the past **decade**. Our hotel has been getting seafood supply from your company for years and has never had any quality issues until the last shipment. The QC Department informs us that the recent shipment we received was not fresh enough for us to **present** to our customers and some items in the shipment were missing. Therefore, we have no choice but to hold on to the payment until the problems get solved.

As you know that the demand for food in our hotel supply is **enormous**, and the **least** we want to see is **crisis** like this that affects both of our businesses. Not to mention that if a food poison problem happens in our hotel, the harm is **unimagina-tive**.

So we are writing today to ask for your cooperation in dealing with the quality issues. We will not be able to make the payment until we receive a GO from the QC Department. Also, the QC Department will recheck all **inventories** from Wild and Fresh Seafood. So if there is any payments that are not due, they will be come **pending** until further notice.

Your usual prompt and positive response of this request would help a great deal at this time.

Yours sincerely,
Daniel Liu
Director of Finance

2013 年 8 月 20 日

Wild and Fresh Seafood Corp.
No.212 Da-An Rd,
Da-An Disctrict, Taipei 106

先生們，

羅森伯格酒店感謝您過去十年的合作。我們的酒店已經向貴公司採買海鮮多年，從未出現過任何品質問題，直到最後一次的發貨。 QC 部門告訴我們，我們近期收到的貨不夠新鮮，使我們無法提供我們的客戶，並發現有一些物品不翼而飛。因此，我們沒有選擇，只能在問題解決前，扣押貨款。

如你所知，我們酒店對糧食供應的需求是巨大的，而我們最不希望看到的是像這樣的危機來影響我們雙方的業務。更何況，如果我們的酒店發生食物中毒的問題，其危害是無法想像的。

所以我們今天的來信是來詢求您處理品質問題的合作。直到我們收到品管部門的進一步通知前，我們將不會進行付款。同時，品管部門將重新檢查貴公司全部庫存。所以如果有任何未付款的款項，目前將全部扣押，直至另行通知。

這時，如同您平時的迅速和積極的回應將會有很大的幫助。

此致，

Daniel Lu
財務總監

在商務往來中偶而會有一些狀況使得付款延遲。此時我們必須提出付款延遲要求。

這篇例文的延遲付款需求是由於對於已收到的貨物品質有疑問，因此將延遲付款。

撰文方面，第一我們應當禮貌性地說明自己與對方的良好關係。這樣的開頭可以和緩內文的批評文字。而後，就應當先說明自己所面對的問題。例如這裡所指的問題是所收到的貨物品質上有問題，必須扣押貨款。如果是我方的問題，例如公司需要更多時間週轉現金等，也應在信件的開頭說明。詳細的說明有助於雙方的溝通與關係的維繫。

最後則是告知在何時可以進行付款，以便對方做出回應。

字彙

1. **Decade**: (n) 十年

   After 3 decades, they still love each other.

   過了 30 年，他們還是深愛彼此。

2. **Present**: (v) 贈送，呈獻

   Kids are going to present the end of year play in school today.

   孩子們今天將在學校呈獻他們的年終舞台劇。

3. **Enormous**: (adj) 巨大的，龐大的

   The trouble he has caused is enormous.

   他所造成的問題是龐大的。

4. **Least**: (adj) 最小的；最少的

   The least we want is to make you come to pick us up.

   我們最不希望的就是讓你來接我們。

5. **Crisis**: (n) 危機；緊急關頭；轉折點

   Japan has been in economic crisis for decades.

   日本的經濟危機已經持續了好幾十年。

6. **Unimaginative**: (adj) 缺乏想像力的；無法想像的

   The harm caused by the storm is unimaginative.

   暴風雨所造成的傷害是無法想像的。

7. **Inventory**: (n) 庫存

   We need to increase the inventory level before the holiday season.

我們必須在節日前增加庫存量。

8. **Pending**: (adj) 懸而未決的；未定的；待定的

The proposal is still pending.

這個提案還未定。

拒絕延遲付款要求 Deny the request of delay on payment

August 21st, 2013
Hotel Rosenberg
313 5th Avenue,
Chung-San District, Taipei 104

Dear Mr. Lu,

Your request for extension of **payment** as presented in your letter of August 20th has been given every **consideration** here.

Unfortunately, our policy with regard to **contractual** agreements does not allow **waiver** of legal rights. You are; therefore, expected to continue to make payments within the limits stated in our agreement. From our understanding, the low quality seafood that got delivered to your facility was caused by the bad temperature control during **transit**. All products that were shipped out from our facility passed our careful **internal** quality check. And since the payment term was set as FOB Shipping Point, the products are no longer our responsibility once they leave our facility. Even so, we are now communicating with the delivery company to come up with the **reparation**. We will make sure that the reparation fulfills your loss and will also confirm that the same problem won't happen again.

Please believe that we do understand the **frustration** you must have, and we sincerely want to solve the problem smoothly. Our company values the business with Hotel Rosenberg very much and will like to continue to work with you after this matter.

Yours sincerely,
Cherry Tsai

中譯

2013 年 8 月 21 日

Rosenberg 酒店

313 5th Avenue,

Chung-San District, Taipei 104

親愛的 Lu 先生，

你在 8 月 20 日的信中所提到延遲付款的要求已經過我們多方考慮。

不幸的是，對於合同協議條款，我們不被允許放棄合法的權益。因此，請您繼續在我們協議規定的時限內支付貨款。就我們的理解，這批被運送到您的地點的低品質海鮮，是由於在運輸過程中的溫度控制不好所造成的。所有由本公司送出的產品均通過我們工廠細心的品質檢查。而且，由於付款期限被設定為 FOB 發運點，因此產品一旦從我們公司寄出，則不再是我們的責任。即便如此，我們現在正在與快遞公司溝通賠償金。我們將確保賠償金可以涵蓋您的損失，也將確認同樣的問題不會再發生。

請相信，我們明白您的沮喪，且我們真誠希望順利解決這個問題。我們公司高度重視與羅森伯格酒店間的生意往來，並希望在這件事情解決以後能繼續與您合作。

此致，

Cherry Tsai

達人提點 II

在收到延遲付款的通知後，當然必須以書面做出回應，尤其是不能同意延遲付款需求時。

但如何能巧妙的拒絕要求呢？同樣的，我們應當在信件的一開頭便表示在某日期收到貴公司的什麼要求並已經過多方的思考。

同意或不同意要求則在第二段的開頭先作說明。Yes or No！如果同意，則在段落中再度說明應當繳款的日期，並請求配合。但若不同意延遲付款的要求，則應該在段落中說明不同意的原因。告知原因是非常重要的，若表明已經經過了努力但還是無法更改原有契約的內容，這可幫助維持兩方間的關係。

在信件最後則是再次說明我方的誠意，以方便以後的再度合作。

A 字彙

1. **Payment**: (n) 支付的款項

He is afraid he doesn't have enough money to make the house payment this month.

他害怕他這個月沒有足夠的錢付房貸。

2. **Consideration**: (n) 考慮

Her application is still under consideration.

她的申請還在被考慮中。

3. **Contractual**: (adj) 契約的

The contractual agreement allows us to deliver parts at the end of the month.

這個契約的同意書讓我們在月底出貨。

4. **Waiver**: (n) 棄權；免責

There is no waiver clause in the contract that we can stop paying the service charge.

合同中沒有免責條款讓我們可以停止付服務費。

5. **Transit**: (v) 運輸，運送 [U]

These goods were damaged in transit.

這些貨物在運輸過程中損壞了。

6. **Internal**: (adj) 內的，內部的

This document is for internal use only.

這份文件是只供內部使用。

7. **Reparation**: (n) 補償；賠償

   She is responsible for all the reparation in the accident.

   在這個意外之中她要負責所有的賠償。

8. **Frustration**: (n) 挫折，失敗，挫敗

   How to stand up from frustration is something everyone needs to learn.

   如何從挫敗中站起來是每個人都必須要學習的。

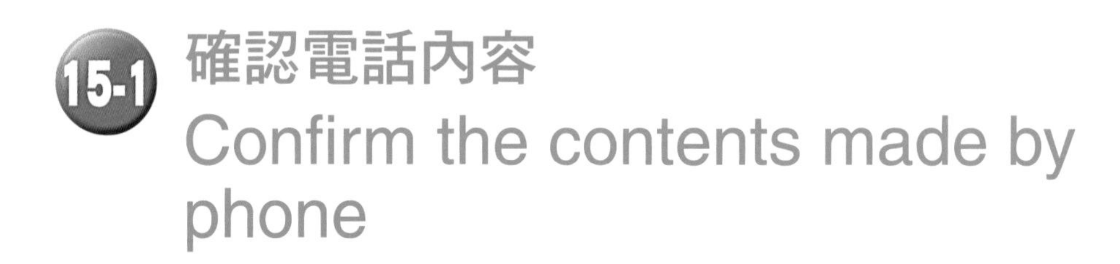

Unit 15 確認 Confirmation

15-1 確認電話內容 Confirm the contents made by phone

November 11th, 2011
Mr. Antonio Flamingo

Dear Mr. Flamingo,

First of all, thank you for being the VIP member of the Mayflower department store. We appreciate your support thru out the year and look forward to **providing** you better service every time.

This letter is to **confirm** the conversation you had with our service center on November 9th and also give you the **follow up** status of the matter. A **complaint** was filed by you to the service center that our staff was not doing a good age control. Over age kids and under age kids were all allowed to go into the playground, which caused over crowded problem and your kid was **injured** because of this matter. As mentioned on the phone, an **immediate** action was taken after you filed the complaint. We now require age check for every person who enters the facility. We also make head count at the gate which allows 20 kids in the area at one time. Each kid is allowed to play for 20 minutes.

We appreciate your call to our service center and let us know our problems. We **value** your opinion always. Therefore, if there are any questions or problems, please do not **hesitate** to give us a call.

We look forward to hearing from you again.

Best regards,
Krystal Darmanic
Customer Service

中譯

2011 年 11 月 11 日
Antonio Flamingo

尊敬的 Flamingo 先生，

首先，感謝您成為五月花百貨的 VIP 會員。我們感謝您每年的支持，並期待著每一次為您提供更好的服務。

這封信是為了確認您曾在 11 月 9 日與我們服務中心的對話，並同時也給您問題的跟進狀態。您向我們的服務中心客訴說我們的工作人員沒有做好年齡控制。年齡過大的孩子和未達年紀標準的孩子都被允許進入遊樂區，造成了擁擠的問題，你的孩子因此而受傷。正如之前電話裡所提到的，接到您的客訴後我們立即採取了行動。現在我們在進入該設施前會先確認年齡。我們還在大門口處控制人數，該地區同時間只能允許 20 個孩子。每個孩子允許玩 20 分鐘。

我們感謝您來電至我們的服務中心，讓我們知道我們的問題。我們始終重視您的意見。因此，如果有任何疑問或問題，請不要猶豫，打電話給我們。

我們期待著再次聽到您的來電。

最好的問候，
Krystal Darmanic
客戶服務

在客服部門我們往往要在電話上與客戶對談，但對談內容因為沒有書面證據，所以容易在之後造成誤會或困擾，所以在電話對談後再寄出一封確認信跟進是比較好的。

跟進書信的撰寫方式，開頭的禮貌文是不能少的。接下來則是說明這封信件是針對哪一天的什麼事件的電話作書面的內容確認及進度跟進。與客戶電話中的內容須先說明，作為確認的用意。如果有需要道歉之處，也應該在這個地方表示。

之後則是說明對於客戶在電話上的問題做出詳細的說明及跟進的進度。

在書信的最後則是再度感謝顧客對店家的支持，並提出意見。“If there are any questions or problems, please do not hesitate to give us a call. We look forward to hearing from you again.”這兩句結束文幾乎是在所有信件回覆中都會看到的。

A 字彙

1. **Provide**: (v) 提供

   This department store provides free delivery service.

   這間百貨公司提供免費送貨服務。

2. **Confirm**: (v) 證實；確定

   The police report confirmed that he is guilty.

   警察的報告證實了他有罪。

3. **Follow up**: 隨訪

   This is the salesman's duty to follow up with customers after the purchase.

   在客戶購物後隨訪是業務的責任。

4. **Complaint**: (v) 抱怨；抗議

   Customer Service Department is in charge of dealing with customers' complaints.

   客服部門負責處理客訴。

5. **Injured**: (adj) 受傷的

   Injured travelers are allowed to board first.

   受傷的旅客可以先行登機。

6. **Immediate**: (adj) 立即的，即刻的

   The immediate answer normally is the most honest answer.

   立即的回答往往是最老實的答案。

7. **Value**: (v) 尊重；重視，珍視

   My father values his friendship with uncle Jim very much.

   我父親非常重視他與 Jim 叔叔的友情。

8. **Hesitate**: (v) 有疑慮，不願意

   He never hesitates to talk about his opinion.

   他從不猶豫說出他的見解。

15-2 澄清誤解 Clarify Misunderstandings

Dear Yuki,

From your email yesterday, it appears there is a misunderstanding of our **intention**. It was the automatic transaxle, not the manual one, that our VP referred to during his visit on June 1st. Please correct any **misinterpretation** here.

We are interested in using your new automatic model in our front-wheel-drive vehicles, since we **anticipate** that demand for manual models will decrease. However, **feasibility** can only be **determined** after receiving drawings and specifications from you, due to width limitations on vehicles here in Taiwan. **Consequently**, we would like you to provide this material as soon as possible.

If your policy does not permit such an accommodation without contract coverage, we will **reluctantly** have to reconsider our plans.

A quick reply would very much **facilitate** planning here.

Sincerely,
Peter Tu

中譯

親愛的 Yuki，

從您昨天的電郵我們瞭解到我們之間似乎有些誤解。我們總裁在 6 月 1 日訪問期間所提到的是自動變速器而不是人工的。請在這裡糾正任何誤解。

由於我們預計手排擋車型的需求將減少，因此我們感興趣的是使用你們的新型自動模式規格在我們的前輪驅動車輛內。然而，由於台灣對車輛寬度的限制，我們只能在收到你們的設計圖及規格後才能評估可行性。因此，我們希望您能盡快提供這種材料。

如果您的規定是在沒有合同的情況下不能提供這種服務，我們則不得不重新考慮我們的計劃。

一個快速的答覆將非常有利於計劃推行。

真誠的，

Peter Tu

達人提點 I

這一封信件的重點是在澄清誤解，並同時告訴對方我方的需求。

首先，我們必須在信件裡提及誤解的所在。這個誤解是在什麼時間、什麼地點、怎麼樣的情況下造成的，並立即要求更改。

在第二段的內文則是再次強調我方的需求以及做出這樣需求的原因。“因為手排擋車的需求將會減少，因此我們要求的是新型的自動模式規格”這說明了若規格錯誤，則會造成爾後很大的問題。

在這封信件裡也提出了“時間限制”及“服務”的要求。換句話說也是一種策略性的書信寫法表示出若對方不同意或無法配合，則我方將更換供應商。這樣的文句也可以幫助促使對方加快回覆。

字彙

1. **Intention**: (n) 意圖，意向，目的
   Harming her is not his intention.
   傷害她不是他的意圖。

2. **Misinterpretation**: (n) 誤解
   Thinking I am going to leave without saying goodbye is a misinterpretation.
   認為我將不告而別是一個誤解。

3. **Anticipate**: (v) 預期，期望；預料
   It is impossible to anticipate someone's thought.
   預料其它人的想法是不可能的。

4. **Feasibility**: (n) 可行性，可能性
   We must always judge the feasibility of success before investing in any business.
   我們在投資任何事業前都應當衡量它成功的可能性。

5. **Determinate**: (adj) 確定的；有限的
   The purchase quantity cannot be determinated until we see the actual sample.
   在未見到實際樣品前購買數量是無法被確定的。

6. **Consequently**: (adv) 結果，因此，必然地
   He lied to his parents and consequently he got grounded.
   他對他的父母說了謊，因此他被禁足了。

新多益

雙篇閱讀篇

NEW TOEIC Double Passages

SAMPLE 1 Business Travel

A Invitation Letter + Itinerary Confirmation

Excelics Microwave Electronics, Inc.
Cor.3rd.St., 3rd Ave., MEPZA, Lapu-Lapu City,
6015 Manila, Philippines
Mr. Robert Rosenberg,
Suite 110, Augustine Drive,
Santa Clara, CA 95054

Dear Mr. Rosenberg,

We are pleased to welcome you to visit our facility for the **annual audit**.

Ms. Emily Chou informed me that you will arrive in Manila on flight CX907 on Wednesday Feburary 11th, at 09:50am. Your personal assitant during your visit Ms. Alanna Wells will meet you at the airport and help you with all your travel needs. If there is any special requirements during your visit, please do not **hesitate** to let her know.

In regard to the hotel and restaurant reservations, please let me know the location and cuisine preferences, and I will arrange accordingly. I look forward to meeting with you.

Sincerely,
Damien Young, General Manager

B Response

Date: Janurary 31st, 2015
From: rrosenberg@remecbb.com
To: d_young@excelics.com
Subject: Vendor facility audit and itinerary comfirmation

Dear Mr. Young,

I appreciate your **proper** response in regard to my visit to your facility next month.

I will arrive in Manila on flight CX907 on Feburary 11th **indeed**. About the hotel resevation, any location downtown with breakfast and high-speed internet is prefered. Dinner reservation is not necessary. I think it will be a good chance for me to explore the city during this trip.

I look forward to meeting with you and your staffs at your facility as well.

Sincerely,
Robert Rosenberg, Vendor Management

Answer the Following Questions

1. Damien's email was sent to:

A Hotel Staff

B Vendor of Excelics Microwave

C Robert's assistant

D Damien's assistant

2 Which of the following is true?

A Robert prefers to have all dinner arrangements pre-set by the vendor.

B Excelics Microwave will assign Emily Chou to be Robert's personal assistant.

C Damien will pick up Mr. Rosenberg at the airport in person.

D The purpose of Mr. Rosenberg's visit is to do the annual audit.

3 When did Robert send his email?

A Janurary 31st

B Not identified

C Feburary 11th

D Feburary 5th

4 What does Allana Wells likely to do after Robert sends the email?

A Make dinner reservation

B Check the arrival time for CX 907.

C Make hotel reservation with highspeed internet and breakfast.

D Call Robert's assistant to confirm flight information.

5. What is Allana Wells' job?

A Robert's personal assistant at Remec BB

B Staff of Excelics Microwave

C Emily Chou's co-worker

D Damien's personal assistant

Answer

1. B　2. D　3. A　4. C　5. B

例文一　出差旅行

A 邀請函及行程確認

親愛的 Rosenberg 先生

我們很高興地歡迎您來參觀我們的工廠及進行年度審核。

Emily Chou 女士告訴我，您將會搭乘 CX907 班機，於二月十一日週三早上九點五十分抵達馬尼拉。您的個人秘書 Alanna Wells 女士將在機場與您碰面，並提供您所有的旅行需求上的協助。如果您訪問期間有任何特殊要求，請儘管告訴她，不要猶豫。

關於酒店和餐廳預訂，請讓我知道您對地點和美食的喜好，我會做適當的安排。我期待與您見面。

誠摯的

Damien Young, 總經理

B 回覆

親愛的 Young 先生，

很感謝您適切地回應我下個月至貴公司的訪問。

我確實會在二月十一日搭乘 CX907 班機抵達馬尼拉。關於酒店的訂房，任何位於市中心並提供早餐和高速互聯網的酒店將為首選。晚餐的預訂是沒有必要的。我認為這對我來說將會是一個遊覽城市的好機會。

我期待著與您和您的員工在您的工廠見面。

誠摯的

Robert Rosenberg

供應商管理

例題中譯

1. Damien 的電郵是寄給誰？

A 酒店人員

B Excelics 微波的供應商

C Robert 的助理

D Damien 的助理

2 以下何者為真？

A Robert 希望供應商可以事前將晚餐都預約好。

B Excelics 微波將指訂 Emily Chou 為 Robert 的個人助理。

C Damien 將會親自到機場接 Rosenberg 先生。

D Rosenberg 先生這次訪問的目的是做年度審核。

3 Robert 是何時寄電郵的？

A 1 月 31 日

B 未說明

C 2 月 11 日

D 2 月 5 日

4 Allana Wells 在 Robert 寄出電郵後將會做甚麼？

A 預訂晚餐

B 確認 CX 907 的抵達時間

C 預訂有高速網路及早餐的酒店

D 致電 Robert 的助理確認航班資訊。

5. Allana Wells 的工作是甚麼？

A Robert 在 Remec BB 的個人助理

B Excelics 微波的工作人員

C Emily Chou 的同事

D Damien 的個人助理

解答

1. B 2. D 3. A 4. C 5. B

解題

1. 在第一封信件裡我們瞭解 Damien 的電郵是寄給 Mr. Rosenberg，而在第二封郵電的署名，我們可以得知 Mr. Rosenberg 是 Excelics 的供應商，因此解答為 B。

2. 在第一封邀請函的開頭 Damien 表明了邀請 Rosenberg 先生到公司做年度審核因此解答為 D。

3. 在第二封回函的日期標明為 1 月 31 日因此解答為 A。

4. 在第二封回函的第二段落中 Robert 表明了希望可以幫忙預定有高速網路及早餐的酒店，因此 Allana Wells 身為助理，接下來的舉動將為 C。

5. 兩篇文中皆沒有說明 Allana Wells 的職稱，因此我們只能知道他是 Excelics 微波的員工，因此解答為 B。

字彙

1. **Annual** (Adj) 年度的，一年一次的

   Most companies have their annual party at the end of the year.

   大部份的公司都在年末舉行年度的派對。

2. **Audit** (v) 檢查，審核，稽核

   Different from most countries, Japanese firms normally have their yearly audit in May.

   與大部份的國家不同，日本的產業一般在五月進行年度稽核。

3. **Hesitate** (v) 躊躇，猶豫，遲疑

   If you need any help, please do not hesitate to ask.

   你如果需要什麼幫助請儘管說，不要猶豫。

4. **Proper** (Adj) 適合的，恰當的

   Students are not allowed to leave campus during class hour without proper authorization.

   學生若沒有適當的授權，其在上課時間是不被允許離開校園的。

5. **Indeed** (Adv) 當然地，真地，確實地

   It is indeed raining in Seattle during this season.

   這個季節在西雅圖當然是在下雨。

SAMPLE 2 Job

A Offer Letter

Mr. Takeshi Sakamoto
3221 Lakeside Drive,
Sunnyvale, CA, 95134

Dear Mr. Sakamoto,

NuSkin Corp. is pleased to offer you the position of Assistant Director, Customer Relations. Your skills and experience will be an ideal fit for our Customer Service Department. As we discussed, your starting date will be April 1st, 2015. The starting salary is $45,000 per year and is paid on a bi-weekly basis. Direct deposit is available. Full family medical coverage will be provided through our company's employee benefit plan and will be effective on May 1st. Dental and optical insurance are also available. NuSkin offers a flexible paid-time off plan which includes vacation, personal, and sick leave. Time off **accrues** at the rate of one day per month for your first year, then increases based on your **tenure** with the company. **Eligibility** for the company retirement plan begins 90 days after your start date. If you choose to accept this job offer, please sign the second copy of this letter and return it to me at your earliest convenience.

When your **acknowledgement** is received, we will send you employee benefit enrollment forms and an employee handbook which details our benefit plans and retirement plan. We look forward to welcoming you to the team. If you have any questions, please let me know, so I can provide any additional information.

Sincerely,
Patricia Camarena
Director, Human Resources
NuSkin Corp.

B Job Acceptance

Takeshi Sakamoto
3221 Lakeside Drive,
Sunnyvale, CA, 95134
+408-386-6732
tsakamoto@amail.com
March 11th, 2015
Ms Patricia Camarena
Director, Human Resources
2110 Technology Rd.
Santa Clara, CA, 95004

Dear Ms Camarena

Thank you for offering me the position of Assistant Director, Customer Relations with NuSkin Corp. I am pleased to accept this offer and look forward to starting employment with your company on April 1st, 2015.

As we discussed, my starting salary will be $45,000 and health and life insurance benefits will be provided after 30 days of employment. There might be some questions about the family coverage which I will try to find a time to discuss with you after joining the team.

Thank you again for giving me this wonderful opportunity. I am eager to join your team and make a positive contribution to the company.

Please find the 2nd signed copy of the **acceptance** letter. If there is any further information or paperwork you need me to complete, please let me know and I will arrange it as soon as possible.

Sincerely,
Takeshi Sakamoto

Answer the Following Questions

1. What kind of job is Takeshi applying for?

- A Human Resource
- B Assistant Director of Sales
- C Manager of Customer Relations
- D Assistant Director of Customer Relations

2 How much will Takeshi get paid each time?

- A $45,000
- B $3,750
- C $1,875
- D $4,500

3 What does Takeshi need to file in in order to accept the job?

- A Passport
- B Employee benefit enrollment forms
- C Family medical insurance
- D Acceptance letter

4 When will Takeshi's insurance start?

- A April 1st
- B May 1st
- C March 11th
- D 90 days after the starting date

5. What does Takeshi want to discuss with Patricia about?

A Retirement plan

B Job resposibility as the assistant director

C Family insurance plan

D Dental and optical insurance plan

Answer

1. D　2. C　3. D　4. B　5. C

例文二 工作

A 聘書

親愛的 Takeshi Sakamoto 先生，

NuSkin 公司很高興將提供您本公司公關部副總監一職。你的技能和經驗將非常適合我們的客戶服務部門。正如我們所討論的，您的起始日期是 2015 年 4 月 1 日，起薪是每年 45,000 美元，並以雙週薪支付。可提供直接存款。我們公司的員工福利計畫將提供完整的家庭醫療保險，並於 5 月 1 日生效。牙科和眼科保險也有提供。NuSkin 也提供了一個靈活的有薪休假計劃，其中包括休假、個人事假和病假。計算方式由開始工作日開始累積，在第一年內為每個月一天，之後將依據您的在職任期增加。退休資格將於任期 90 天後開始計算。如果您選擇接受這個工作機會，請在您方便時盡快回簽這封信的第二個副本給我。

當收到您的確認，我們將寄給您員工福利登記表和員工手冊，詳細介紹我們的福利計劃和退休計劃。我們期待著歡迎您加入我們的團隊。如果您有任何疑問請讓我知道，所以我可以提供任何額外資訊。

真誠的，

Patricia Camarena

人力資源總監

NuSKin 公司

B 接受工作邀約

Takeshi Sakamoto

3221 Lakeside Drive,

Sunnyvale, CA, 95134

+408-386-6732

tsakamoto@amail.com

2015 年 3 月 11 日

Ms Patricia Camarena

Patricia Camarena

人力資源部總監

2110 Technology Rd.

Santa Clara, CA, 95004

尊敬的 Camarena 女士

謝謝貴公司提供我公關部副總監一職。本人非常開心地接受這個職務，並期待在 2015 年 4 月 1 日開始任職。

正如我們討論的，我的起薪為 45,000 美元，健康和生命保險福利將於就業 30 天後開始提供。對於家庭覆蓋方面，我可能有一些疑問，我會在加入貴團隊後盡量找一個時間與您討論。

再次感謝您給我這個美好的機會。我渴望加入您的團隊，並給公司帶來積極的貢獻。

請查收錄取通知書的簽署副本。如果有需要我提供任何進一步的信息或資料，請讓我知道，我會盡快安排。

真誠，

Takeshi Sakamoto

例題中譯

1. Takeshi 是應徵甚麼樣的工作？

A 人力資源

B 銷售部副總監

C 公關部主任

D 公關部副總監

2. Takeshi 每次所拿的酬勞有多少？

A $45,000

B $3,750

C $1,875

D $4,500

3. Takeshi 需要繳交甚麼資料以表示接受這份工作？

A 護照

B 員工福利登記表

C 家庭醫療保險

D 接受信

4. Takeshi 的保險將在何時開始？

A 4 月 1 日

B 5 月 1 日

C 3 月 11 日

D 開始工作 90 天後

5. Takeshi 希望和 Patricia 討論甚麼？

A 退休計畫

B 副總監的工作內容

C 家庭保險計畫

D 牙科及眼科保險計畫

解答

1. D 2. C 3. D 4. B 5. C

解題

1. 在第一封的聘書的第一段裡提到 NuSkin 給與 Takeshi 先生的職務是公關部副總監，因此解答為 D。

2. 在第一封聘書的第一段中提到 Takeshi 先生的年薪為 45,000 美元，並為雙週薪制，因此每次所拿的金額應該為 45,000 元 /12 個月 /2 週 =1,875 元。解答為 C。

3. 在第一封聘書最後有提到請 Takeshi 先生回簽聘書的副本，以作為接受這份工作的證據，即為接受信，因此解答為 D。

4. 在第一封聘書中段說明保險將在 5 月 1 日生效，因此解答為 B。

5. 在第二封回函的第二段提到 Takeshi 對於家庭保險覆蓋方面有疑問，並希望與 Patricia 討論，因此解答為 C。

字彙

1. **Accrue** (v) 增大，增多
   Interest accrues very slowly on a bank now due to a low interest rate.
   由於低利率，銀行帳戶上的利息增加的非常慢。

2. **Tenture** (n) 保有權，任期
   He remains responsible and aggressive during his tenture as the president of the company.
   他在擔任公司總監的任期內都維持責任且積極的態度。

3. **Eligibility** (n) 必要條件，資格
   Eligibility for same sex marriage was approved by the US government recently.
   同性婚姻的資格最近被美國政府所通過。

4. **Acknowledgement** (n) 回函，謝函
   An order acknowledgement is always required to be sent by the seller after the client receives order from a customer.
   當賣家接獲客戶的訂單時賣家皆須寄出訂單回函。

5. **Acceptance** (n) 接受，收受，承認
   It is harder and harder these days to get acceptance by global companies if you do not have some kind of remarkable skills.
   現在若你沒有很出眾的技能，則越來越難被世界級的公司所接受。

SAMPLE 3 Advertisement

A Workshop Advertisement

Recognize And Successfully Perform Key Duties of A Fixed Asset Manager

Your immediate **takeaway**
• Identify the contents of the fixed asset database
• Create and run effective controls on fixed asset data
• Evaluate how to get the best return for fixed assets

The nature and importance of the fixed asset manager's job is frequently misunderstood and underestimated, but the role is a **crucial** one. Anyone in this position must keep the contents of the fixed asset database accurate and current, provide correct information to stakeholders, and have the skills to get the maximum return on fixed assets. Knowing what data is needed for timely and accurate reporting in financial and tax statements, insurance valuation and recovery, and maintenance of assets is also **essential**. This course is designed to provide you with the tools, knowledge, and confidence to successfully perform all of these duties — and much more.

How You Will Benefit

Employ management techniques essential to successful fixed asset managers

Identify the contents of the fixed asset database and recognize its importance

Create and run effective controls on your organization's fixed asset data

Effectively evaluate fixed asset software

Analyze how to conduct fixed asset inventory and reconciliation

Recognize how fixed assets impact tax returns

Acquire knowledge and tools to help ensure optimum return from fixed asset investments

Schedule

We have 5 scheduled sessions located nationwide starting between 7/20/2015 - 12/20/2015

Date	Location	Duration
Jul 20, 2015-Jul 22, 2015	New York, NY	3 days
Sep 2, 2015-Sep 4, 2015	Chicago, IL	3 days
Oct 15, 2015-Oct 17, 2015	San Francisco, CA	3 days
Nov 9, 2015-Nov 11, 2015	Seattle, WA	3 days
Dec 15, 2015-Dec 17, 2015	New York, NY	3 days

Registering more than 4 people, please call 1-800-555-2241.

B Workshop Inquiry For More Than 4 People

To: Catherine Huang (chuang@improve.com)
From: Kelvin Roberts
Date: April 11th, 2015
Subject: Private Workshops

Dear Catherine,

I am interested in learning more about the private workshops your company can offer for our firm. Currently, our staff is all required to possess all basic stock licenses, but we found it is not enough to help our clients maximize their assets and is not a guarantee of exceptional performance. As the ways of investments are going digital and global, there still are some important areas that the analysts are lacking in. We hope that through attending your workshops, the brokers will be able to supplement their knowledge with concrete knowledge in advanced areas, such as e-commerce, **globalization**, and such is able to more accurately advice our clients where to invest into the money market. If possible, please send me back a detailed curriculum plan and a tuition form. These papers will have to be submitted to the president of the company and the finance department Sothat I can authorize using company dollars to pay for these classes. Thank you.

Sincerely,

Kelvin Roberts

JB Investments

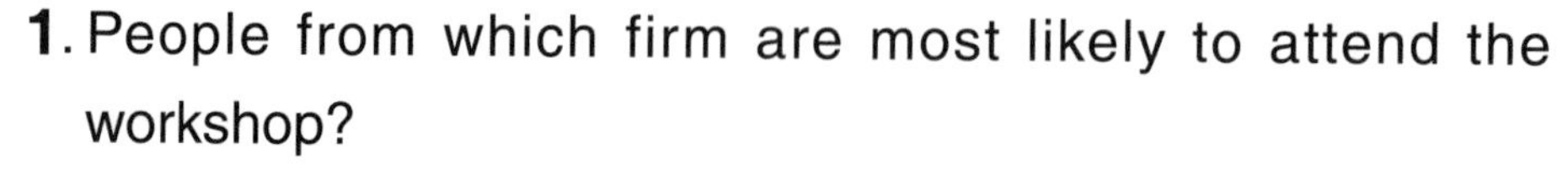

1. People from which firm are most likely to attend the workshop?

A Semiconductor

B Non-profit

C Network

D Finance

2. What will you not benefit from the workshop?

A Run effective controls on fixed asset data

B Analyze how to conduct fixed asset inventory and reconciliation

C Gain personal skills to ensure optimum return from fixed asset investments

D Increase tax return by conducting to fixed asset

3. Where will the workshop be taken place in August?

A Chicago

B San Francisco

C Not identified

D Las Vegas

4. What does Kelvin expect his staff to learn from the workshop?

A Localization

B Stock Licenses

C E-commerce

D Optimum return from investment

5. What kind of company is Improve.com?

A Investment Company

B Stock Broker

C Consulting Company

D Advertisement Copmany

Answer

1. D 2. D 3. C 4. C 5. C

例文三 廣告

A 研討會廣告

認識並成功地執行一個固定的資產管理公司的關鍵任務

您的直接收穫

- 確定固定資產數據的內容
- 創建並有效控制及運行固定資產
- 評估如何獲得固定資產的最好利益回報

固定資產經理工作的性質和重要性經常被誤解和低估，但這是一個關鍵的角色。任何人在這個位置上必須保持固定資產數據的內容準確度和更新度，進而提供利益關係者正確的信息，並有能力獲得固定資產最大的回報。知道在進行財務和稅務報表時需要提供的報表，保險的估值和回收，以及維持資產同時，準確地報告數據也是必不可少的。本課程旨在為您提供工具、知識和信心，成功地完成所有任務，甚至更多。

您將如何受益

- 確定固定資產數據庫的內容，並認識到它的重要性
- 創建並有效控制運行貴組織的固定資產
- 有效評估固定資產的軟體
- 分析如何進行固定資產盤點與對帳
- 認識固定資產如何影響納稅申報
- 獲取知識和工具，以幫助確保從固定資產投資的最佳回報

時間表

由 2015 年 7 月 20 至 2015 年 12 月 20 日將在全國有 5 個場次。

日期	地點	為期
2015 年 7 月 20 日 -22 日	紐約州紐約市	3 天
2015 年 9 月 2 日 -4 日	伊利諾伊州芝加哥	3 天
2015 年 10 月 15 日 -17 日	加州舊金山	3 天
2015 年 11 月 9 日 -11 日	華盛頓州西雅圖	3 天
2015 年 12 月 15 日 -17 日	紐約州紐約	3 天

若超過 4 人註冊，請致電 1-800-555-2241。

B 關於增加四人以上的研討會提問

至：Catherine Huang（chuang@improve.com）
來自：Kelvin Roberts
日期：2015 年 4 月 11 日
主題：私人研討會

親愛的 Catherine，

本人有興趣了解更多關於貴公司可提供本公司的私人研討會。目前，我們的工作人員都必須具備基本的所有股票許可證，但我們發現它是不足以幫助我們的客戶大額度的提升他們的資產，也不是業績的保證。由於投資的方式將會導向數碼化和全球化，這是我們的分析師仍缺乏的一些重要領域。我們希望通過參加您的研討會，券商將能夠具體補充先進的知識領域，如電子商務、全球化等，如此能夠更準確地建議我們的客戶投資在正確的貨幣市場。如果可以，請回覆我一個詳細的課程計劃和學費表。這些文件將提交給公司總裁和金融部門，本人方可使用公司授權的款項支付這筆費用。謝謝。

真誠的，
Kelvin Roberts
JB 投資

例題中譯

1. 甚麼產業的人最有可能參加這個研討會？

A 半導體

B 非營利

C 網路

D 金融

2. 參加此研討會，不會得到什麼好處？

A 有效控制固定資產

B 分析如何進行固定資產的盤點與對帳

C 獲得能確保從固定資產投資中得到最佳回報的技能

D 通過開展固定資產來增加退稅

3. 此研討會八月會在哪裡舉行？

A 芝加哥

B 舊金山

C 未說明

D 拉斯維加斯

4. Kelvin 期待他的員工在此研討會中學到甚麼？

A 本土化

B 股票許可證

C 電子商務

D 投資的最大回收

5. Improve.com 是甚麼樣的公司？

A 投資公司

B 股票經理人

C 諮詢公司

D 廣告公司

解答

1. D　2. D　3. C　4. C　5. C

解題

1. 此文是有關資產管理的研討會，所以解答為D。
2. 在第一篇廣告的“您將如何收益”部份只有提到“如何影響納稅申報”，並沒有提到增加退稅，因此解答為D。
3. 在第一篇廣告中的行事曆中並未提到 8 月的行程，因此答案為C。
4. 在第二篇的信件中，Kelvin 提到希望員工可以學習全球化及電子商務，因此解答為C。
5. 在這兩篇文章中我們可以得知 Improve.com 專辦各式研討會並可以依照客戶的需求擬定課程，因此 Improve.com 應該為諮詢公司，解答為C。

SAMPLE 4 Event Notice

A Event Notice

With 100+ performances featuring jazz, blues, R&B, funk, New Orleans, zydeco and more, San Jose Jazz Festival, which began in 1990, has become one of the nation's most important music **festivals**. Our jazz beyond stage shows Young jazz artists' push of the frontier of the art form, an art form **influenced by** the hip-hop, R&B, neo-soul, and electronic music.

Summer Fest has the twelve indoor and outdoor stages that provide a **variety** of environments as well as music. The Fest's centerpiece is the Kaiser Permanente Main Stage located in the Plaza de Cesar Chavez Park, fronting the Fairmont San Jose. Our event is a feast for dancers – you'll want to get up and move to many of the acts on the Main Stage, Blues Stage and Big Easy Stage, and there are hundreds of couples enjoying world-class live salsa from noon 'til night at the nearby Salsa Stage. Get 50% off tickets for over 100 performances, 12 stages held over 3 days from August 5th – 7th. Adult over 21 years of age can also join Happy Hour from 5p.m. to 7p.m.

Visit http://summerfest.sanjosejazz.org/ for show schedules and performers' index. 100 discount tickets are available today and may be obtained by emailing summerfest@

sanjosejazz.org. Event volunteer opportunities are also available. Email volunteer@sanjosejazz.org for details.

B Request for tickets

To: volunteer@sanjosejazz.org
From: lindsey.smith@imail.com
Date: June 22nd, 2015
Subject: Volunteer Opportunities

I am writing this in response to your Summer Fest ad in the San Jose Mercury News. We are 6 undergraduates in San Jose State University who are **passionate** about Jazz and have been attending the Summer Fest yearly. For the past 2 years, we had missed the opportunity to **volunteer** in the event and are hoping that we can get a chance to help out this year. Please kindly advise which workshop we can volunteer and how to sign up. We are up to any kind of jobs for all 3 days.

Please also kindly send me 10 discount tickets for my family and friends who are not able to volunteer but would like to attend the event during those 3 days. My address and contact information is as following.

Lindsey Smith
275 E San Fernando St, Apt #2
San Jose, CA 95112
Mobile: 408-774-3658

Answer the Following Questions

1. What kind of event is this?

A Beer Festival

B Music Festival

C Salsa Dance Contest

D Adults and Kids Playground

2. Where is the main stange?

A The Plaza de Cesar Chavez Park

B Fairmont San Jose

C Kaiser Permanente

D Inside the university

3. Where can you get discount ticket?

A By visiting the San Jose Jazz Organization

B By emailing volunteer@sanjosejazz.org

C By emailing summerfest@sanjosejazz.org

D By visiting http://summerfest.sanjosejazz.org/

4. What does Lindsey Smith do?

A A journalist of San Jose Mercury News

B A volunteer of the Jazz Festival

C A staff of the Jazz organization

D A student of San Jost State University

5. Which is not in Lindsey's wishlist?

A Discount tickets

B Attend the event

C A job

D Voltunteer opportunity

Answer

1. B　2. A　3. C　4. D　5. C

例文四　活動通知

A 活動通知

具有超過百個以爵士、藍調、R&B、放克、紐奧良、柴迪科為特色等的表演，源於 1990 年代的聖荷西爵士音樂節，已成為國家最重要的音樂節之一。我們的「超越爵士舞台」呈現出年輕藝術家推動藝術形式的界限，且受到嘻哈、R&B、新靈魂音樂和電子音樂的影響。

夏季音樂節有著十二個室內和室外的舞台提供多種環境以及音樂。音樂節的主舞台是 Kaiser Permante，它將設在 Cesar Chavez 公園廣場，面向費爾蒙聖荷西。我們的活動是一場舞者的盛宴 - 你會想要站起來移動到許多的舞台，包括主舞台，藍調舞台和紐澳良舞台等，那裡並會有數百對夫妻在附近的莎莎舞台，從中午到晚上享受世界級的現場莎莎舞。

現在就獲得 50%折扣的門票！從 8 月 5 日至 7 日起的三天，參觀 12 個舞台，超過 100 個表演。超過 21 歲的成人還可以參加從下午 5 點到下午 7 點的歡樂時光。

請至 http://summerfest.sanjosejazz.org/ 參閱表演時間表和表演者索引。 今日起可通過電子郵件 summerfest@sanjosejazz.org 索取折扣門票，共 100 張。並有擔任大會志工的機會。志願者請來函 volunteer@sanjosejazz.org 了解詳細信息。

B 票券請求

至：volunteer@sanjosejazz.org
來自：lindsey.smith@imail.com
日期：2015 年 6 月 22 日
主題：志願者機會

這封郵件是要回應您在聖荷西水星新聞所登的夏季音樂節的廣告。

我們六人就讀聖荷西大學熱衷於爵士樂，並每年參與夏季音樂節的大學部學生。在過去的 2 年，我們錯過當活動志工的機會，所以希望我們今年能有機會幫忙。請讓我們知道是否有志願者說明會並了解如何註冊。我們願意在 3 天內做任何的工作。

也請好心的寄給我 10 張折價入場券給我無法做志工但也想在這 3 天內參加活動的家人和朋友。我的地址和聯繫信息如下。

Lindsey·Smith
275Ë 聖費爾南多街，公寓 # 2
聖荷西，加利福尼亞 95112
手機：408-774-3658

例題中譯

1. 這是甚麼樣的活動？

Ⓐ 啤酒節

Ⓑ 音樂節

Ⓒ 莎莎舞競賽

Ⓓ 大人與小孩的遊樂場

2. 主舞台在哪裡？

Ⓐ Cesar Chavez 公園廣場

Ⓑ 聖荷西費洛蒙酒店

Ⓒ Kaiser Permanente

Ⓓ 大學內

3. 你可以從哪裡得到折扣門票？

Ⓐ 通過訪問聖荷西爵士組織

Ⓑ 通過電郵 volunteer@sanjosejazz.org

Ⓒ 通過電郵 summerfest@sanjosejazz.org

Ⓓ 通過參閱 http://summerfest.sanjosejazz.org/

4. Lindsey Smith 的工作是甚麼？

Ⓐ 聖荷西水星新聞的記者

Ⓑ 爵士節的義工

Ⓒ 爵士組織的成員

Ⓓ 聖荷西州立大學的學生

5. 何項不是 Lindsey Smith 想得到的？

A 折扣門票

B 參加活動

C 一份工作

D 當義工的機會

解答

1. B　2. A　3. C　4. D　5. C

解題

1. 在活動通知的開頭並提到這是爵士音樂節，因此解答為 B 。
2. 在活動通知的第二個段落裡提到，Kaiser Permanente 主舞台是位在 Cesar Chavez 公園裡，因此解答為 A 。
3. 活動通知的第三段裡說明，通過電子郵件 summerfest@sanjosejazz.org 可索取折扣門票，因此答案為 C 。
4. 在第二篇的信件中，Lindsey Smith 說明自己及朋友是一組六人就讀聖荷西大學部的學生，因此解答為 D 。
5. 在第二篇的信件中，Lindsey Smith 表明了希望能當義工的機會，並要求折扣門票以參加活動，但未提及成為正式員工的需求，因此解答為 C 。

字彙

1. **Festival** (n) 定期在某地舉行的節日，例如：紅酒節，美食節等。
   Beer Festival in Germany is legendary and always attracts thousands of people.
   啤酒節在德國是非常聞名的且經常吸引成千上萬的人。

2. **Influenced…by** 被…所影響
   Parents should always know who their kids are hanging out with because kids are easiliy influenced by their surroundings.
   父母應該隨時知道他們的孩子跟誰在一起，因為孩童很容易被周圍所影響。

3. **Variety** (n) 多樣化，豐富多彩
   Key point of success for online stores is the variety.
   線上商城的成功之道就是多樣性。

4. **Passionate** (adj) 熱切的，強烈的
   It is important to find a job that you are passionate about in order to show your best performance.
   找一份你有熱忱的工作是很重要的，這樣才能秀出你最好的表現。

5. **Volunteer** (n) 志願者 ; (v) 自高奮勇的，自願效勞的
   David volunteered to drive his roommates to the party tonight.
   David 自願載他的室友們去今晚的派對。

SAMPLE 5 Food Critic

A Critic

I'm always **down for** some sweets. I think the chase for the perfect dessert, such as pie is an ongoing **quest**. To be honest, I had never heard of this place. I don't exactly get to that side of town that often and because of the location, and I might have missed it on a casual drive by kind of day.

Surprisingly, the food at Steven's Snack Shop appears simple, but the flavors have remarkable depth. Steven's is known best for its range of fruit pies, including apple, blueberry, rhubarb, and peach, and the reputation is well deserved. The fruits are harvested from local organic farmers, and the pies are served warm with fresh homemade vanilla ice cream **crafted with** real vanilla beans. A host of other goodies, including brownies, cookies, and candies, line the shelves, offering something for the whole family. The shop is both clean and tidy—despite having limited space and more customers than it can fit inside at one time—and all the staff members there greeted my companions and me with smiles as warm as the pies they serve there. I would **thoroughly** recommend Steven's Snack Shop to anyone with a sweet tooth. I know I'll certainly be returning soon.

B Response

Foodforgood.com
300 S. First Street,
Chicago, IL 60602

Dear Foodforgood.com,

I read your article regarding to the Steven's Snack Shop and am writing you this mail to tell you that I couldn't agree more. I was born and raised in this town for 36 years, and this is how long I have been going to Steven's Snack Shop. Steven's Snack Shop is not only a place that offers wonderful food, but also gives tons of kids who grow up in this town wonderful memories. Even though, we are seeing many chain stores, such as Starbucks expanding their power in every single corner of Chicago, and are doing a marvelous job by offering their customers' quick and quality food and drinks, I would say that we still need local stores (like Steven's Snack Shop) which show the true color of our town.

As great as your critic was, there was one **distinguished** feature you might have left out in your article which the fondant cake Steven's Snack Shop creates. Back in the days, we used to call Steven the Willy Wonka of the cake world. He had the purest heart and the best creativity, which made the kids' dream cakes come true. We could order any kind of imaginary cake, and Steven always delivered. For years, even though Steven doesn't make cake by himself anymore, staffs in Steven's Snack Shop kept the tradition and continued with the kid's birthday cake orders. I am sure that no one can overcome the happiness they bring to the city.

Again, I appreciate that you spend the time visiting the local store here in Chicago, and had a wonderful experience. I wish your business the best and look forward to reading more food critics from your sites.

Sincerely,
Michael Davis

Answer the Following Questions

1. Which of the following statements best summarizes the above news story?

(A) Steven's Snack Shop sells different kinds of snacks that invented by Willy Wonka.

(B) Steven's Snack Shop is a local friendly store that has been loved for years.

(C) Steven's Snack Shop serves full meal from breakfast to dinner.

(D) Steven's Snack Shop is a well-known store that the critic knows about this place before his visit.

2. What was NOT mentioned by the critic?

(A) Steven's cake shop is also famous for its custom birthday cake.

(B) The location of the shop is hard to be noticed by non-local people.

(C) The critic describes the food at Steven's has remarkable depth.

(D) The critc will for sure return back to the store in the future.

3. What is the purpose of Michael Davis' letter?

(A) Disagree with the critic's point of view.

(B) Asking for detail information.

(C) Appreciate the article and add on extra information about the store.

(D) Introduce another local store that he thinks is even better.

4. Which of the following is NOT true?

A Steven's cake shop uses can fruit.

B The atmosphere at the store is very nice and friendly.

C Custom birthday cake fullfilled a lot of local kid's childhood memories.

D Steven's cake shop prefers to staying as a small local store instead of becoming a chain store.

5. How long has Steven's cake shop been in business?

A 36 years

B 10 years

C Over 36 years

D Less than 36 years

Answer

1. B 2. A 3. C 4. A 5. C

例文五 美食評論

A 評論

我總是喜歡來點甜食。我想，追逐完美的甜點如甜派等是一個無止境的探索。說實話，我從來沒有聽說過這個地方。老實說我不常到城鎮的那一頭，並由於地理位置的關係，在偶然的機會裡，我可能錯過了這個地方。

出人意料的是，史蒂芬的小吃鋪的食物雖然顯得簡單，但味道確有顯著的深度。史蒂芬最受稱讚的是它一系列的水果派，包括蘋果、藍莓、大黃、桃仁等，它的聲譽是當之無愧的。使用的水果是由當地有機農民採收的。他們的派是溫熱的，並與真正香草莢所製造的香草冰淇淋一起供應。店裡還有許多其他好吃的東西，包括布朗尼、餅乾、糖果等，架上的產品足以提供一整個家庭的需求。這家商店既乾淨且整潔，儘管空間有限且客人總是多於可容納的範圍，所有的工作人員皆用和他們派一樣溫暖的笑臉迎接我的同伴和我。我會全力推薦史蒂芬的小吃鋪給任何愛吃甜食的人。我知道，我以後一定會很快再回到這裡。

B 回覆

Foodforgood.com
300 S. First Street,
Chicago, IL 60602

親愛的 Foodforgood.com，

我閱讀了您寫關於史蒂芬點心鋪的文章，因此寫這封信來告訴您，我非常同意您的看法。我出生並成長在這個城鎮 36 年，這也是我來史蒂芬點心鋪多久的時間。史蒂芬點心鋪不僅是一個提供美妙食物的地方，也是給在這個城鎮成長的孩子美好回憶的地方。雖然我們看到很多如星巴克的連鎖店伸展了他們的權力在芝加哥的每一個角落，也提供其客戶快速、優質的食品和飲料，展現了優秀的工作表現，但我必須說，我們還是需要像史蒂芬點心鋪這樣當地的商店來說明我們城市的真實色彩。

雖然您的評論非常好，您可能漏掉了史地芬點心鋪做翻糖蛋糕的特色。過去，我們習慣稱史蒂芬為蛋糕界的威利旺卡。他有最純淨的心和最好的創意，讓孩子們的夢想蛋糕成真。我們可以訂購任何一種假想的蛋糕，史蒂芬始終交付。多年來，儘管史蒂芬並不由他自己做蛋糕了，在史蒂芬點心鋪的工作人員維持了傳統，繼續接受孩子的生日蛋糕訂單。我相信沒有人可以戰勝他們給城市帶來的快樂。

再次地謝謝您花時間在這裡參觀芝加哥當地的商店，並有一個美好的經歷。祝生意興隆並期待可以從您的網站讀到更多的美食評論。

真誠的，

Michael Davis

例題中譯

1. 下列何者總結了上述的新聞？

Ⓐ Steven 蛋糕屋提供 Willy Wonka 所做的各種點心。

Ⓑ Steven 蛋糕屋是個被當地愛戴多年的當地商店。

Ⓒ Steven 蛋糕屋提供早餐到晚餐的全天候餐點。

Ⓓ Steven 蛋糕屋是大家所熟知的商店。評論家在拜訪前就知道這個地方。

2. 下列何者未被闡述在評論中？

Ⓐ Steven 蛋糕屋是因為客製化生日蛋糕聞名。

Ⓑ 小吃店的地點很難被非當地人注意到。

Ⓒ 評論家闡述 Steven's 的食物很有深度。

Ⓓ 評論家以後會再光顧這家店。

3. Michael Davis 寫這封信的原因是什麼？

Ⓐ 不同意評論家的看法。

Ⓑ 詢問更多資訊。

Ⓒ 表示對文章的喜愛並追加更多商店資訊。

Ⓓ 介紹另一家他認為更好的本地商店。

4. 以下何者有誤？

Ⓐ Steven 蛋糕屋使用罐頭水果。

Ⓑ 商店的氣氛非常好及友善。

Ⓒ 客製化生日蛋糕實現了很多當地孩子的回憶。

Ⓓ Steven 蛋糕屋希望維持當地小店而不想成為連鎖商店。

5. Steven 蛋糕屋已經在業界幾年了？

A 36 年

B 10 年

C 超過 36 年

D 低於 36 年

解答

1. B　　2. A　　3. C　　4. D　　5. C

解題

1. 由於 Steve 蛋糕屋結合了 A,C,D 等的好處，因此在當地被大家所喜愛，所以答案為B。

2. 客製化生日蛋糕是由回信給評論家的 Michael Davis 所提出，因此答案為A。

3. 在 Michael Davis 的回覆中闡述了同意並喜愛評論家的看法，並追加了客製化生日蛋糕的資訊，因此 答案為C。

4. 在第一篇評論的第二個段落裡提到，Steve 蛋糕屋只使用新鮮水果，因此解答為A。

5. 在第二篇的回信中 Michael David 提到他在這個城市生活了 36 年，也是他光顧 Steve 蛋糕屋的時間，因此 Steve 蛋糕屋已經存在至少 36 年，所以解答為C。

字彙

1.**Down for** 報名加入

What restaurant are you down for your birthday celebration?

今年生日你想去哪家餐廳？

2. **Quest** (n) 探索

The continuing quest for a cure for cancer is a long journey.

持續探索治療癌症的方法是一個很長的旅程。

3. **Variety** (n) 多樣化，豐富多彩

Key point of success for online stores is the variety.

線上商城的成功之道就是多樣性。

4.**Thoroughly** (adv) 徹底地

After a hard day of work, I feel thoroughly tired.

經過一天辛苦的工作，我覺得累極了。

5. **Distinguished** (adj) 卓越的，著名的

Each city has its distinguished sites that attract travelers to visit.

每個城市都有其著名並吸引觀光客的景點。

SAMPLE 6 Contract Manufacturer Selection

A Comparison email

To: Neil Stevens
From: Sarah McCaughan
Date: September 18th, 2009
Subject: Contractor Selection

Dear Mr. Steven,

The two factories I mentioned in the meeting are located in Changdu, China, and Cebu, Philippines. Both factories have similar **capacity** of 5,000 units per day and run **around the clock** six days a week with one assembly line restarting per week. 2 major differences between these 2 factories are the restart time and worker's language ability. The assembly line restart takes only eight hours in Changdu and is **significantly** quicker than the Cebu assembly line which takes up to twenty-four hours to restart. On the other hand, most workers in Changdu do not speak English, which will be difficult for our staff to communicate with, where over 90% workers in Cebu speak decent English and there will be no language **barrier** at all. Both factories are very efficient and shipments begin almost immediately after the first batch rolls off the assembly line. Both factories will be able to supply all our North American stores to the agreed upon inventory levels within six weeks. Both factories are pretty much equal in terms of capacity, but in terms of efficiency, I think Changdu might be better. On the other hand, it will be a great help if we don't face any communication problems during production. Please advice which factory we should choose.

Best Regards,

Sarah McCaughan

B Response

To: Sarah McCaughan
From: Neil Stevens
Subject: Contractor Selection

Dear Ms. McCaughan,

I'd like to get more information on both factories since neither factory we have worked with before. I have heard good and bad comments about both places, and I do place an **emphasis** on reliable service. Currently, my thought is more toward Cebu since I think good communication skills are extremely helpful when it comes to production. However, before any decision can be made, we need to make sure that both factories can not only handle our order but also handle it for a significant period of time. We are looking for a factory to produce 500,000 units with the possibility of adding another 250,000 units since our business plan is to expand our territory to other continents. We need to audit both factories personally and meet with plant managers. We also need to set up a formal agreement that our company gets the first priority when it comes to high production seasons. We want our products get manufactured and shipped before others. I believe this condition will drag the unit price up significantly from both factories, and will become one of the important judge points for us. Please arrange the audits and arrange the travel time with my personal assistant.

Sincerely,
Neil Stevens
CEO

Answer the Following Questions

1. How many hours per day does the factory in China run?

A 8 hours

B 12 hours

C 24 hours

D Not mentioned in the article

2. Which of the following statement is true?

A All workers speak fluent English in both factories.

B It is better to choose the factory in Cebu if the main concern is the restart time.

C Neil has decided to use the factory in China.

D Neil demands to negotiate the terms and condition with both places.

3. Which is NOT beneficial to use the factory in Cebu?

A Fast shipment

B Good communication skill

C Fast restart time

D Maintain inventory level

4. What will Neil negotiate with both factories ?

A Turnover time

B Inventory level

C First prority at production

D Production capability

5. Neil will most likely to choose the factory in China if which statement is true?

A If all workers in China can speak English.

B It is not mentioned in the email.

C If factory in China increase the production rate.

D If factory in China speed up the retart time.

Answer

1. C　2. D　3. C　4. C　5. B

例文六 合約製造商選擇

A 比較信件

此致：Neil Stevens
來自：Sarah McCaughan
日期：2009 年 9 月 18 日
主題：承包商的選擇

親愛的史蒂芬先生，

這兩個我在會議上提到的工廠分別位於中國成都和菲律賓宿霧。這兩家工廠每天可以生產 5000 件，每星期有六天實行 24 小時運轉。這 2 家工廠的 2 個主要區別是在於重啟時間的不同和工人的語言能力。成都的重啟時間只需 8 小時，比宿霧的重啟時間 24 小時明顯更快速。另一方面來說，大多數在成都的工人都不會說英語，這將造成我們員工溝通上的困難。在宿霧，超過 90%的工人可以說流利的英文，不會有語言障礙。這兩家工廠的效率非常高，配裝生產線開始運轉後就可以馬上裝運，六週內就可以供應北美商店議定的庫存量。這兩個工廠都在產量方面幾乎相等，但在效率方面，我認為成都可能會更好。在另一方面，如果我們在生產過程中可以減少溝通上的問題，這將是一個很大的幫助。請指教我們應該選擇哪個工廠。

最好的問候，
Sarah McCaughan

B 回覆

此致：Sarah McCaughan
來自：Neil Stevens
主題：承包商的選擇

尊敬的 McCaughan 女士，

因為以往我們一直沒有和這兩家工廠有過交流，我希望可以得到更詳細的信息。兩家工廠我都聽說過好的和壞的評論。我將重點放在可靠的服務。目前，我的想法是比較朝向宿霧，因為我認為當涉及到生產時，良好的溝通技巧是非常有用的。然而，在做任何決定之前，我們需要確保兩個工廠不僅可以處理我們現在的訂單，而且可以長時間處理我們的訂單。我們正在尋找一個產能 50 萬台，並可再增加 25 萬產能的工廠，因為我們有計劃將我們的版圖拓展到其它大陸。我們必須親自審核工廠並與工廠經理見面。我們還需要在正式協議中要求在旺季時我們公司能得到第一優先。我們希望我們的產品第一優先生產並在他人之前發貨。我相信這種要求將會拖動兩個工廠顯著的單元價格上漲，這也將成為我們重要的判斷點之一。請安排審核並與我的個人助理安排出差時間。

真誠的，

Neil Stevens

首席執行官

例題中譯

1. 中國的工廠每天運行幾小時？

A 8 小時

B 12 小時

C 24 小時

D 文章中未提及

2. 以下何者為真？

A 兩個工廠的所有員工皆會說流利的英文。

B 若在意的是重啟時間，則應當選擇宿霧的工廠。

C Neil 決定使用中國的工廠。

D Neil 要求與兩家工廠談判條款與條件。

3. 以下何者不為使用宿霧工場的好處？

A 快速出貨

B 好的溝通技巧

C 快速的重啟時間

D 維持庫存水平

4. Neil 將會與兩家工廠談判什麼？

A 週轉時間

B 庫存量

C 第一順位生產

D 生產能力

5. 若下列何者為真，Neil 則會選擇中國的工廠？

Ⓐ 若所有員工皆會說英文。

Ⓑ 郵電中並未提及。

Ⓒ 若中國工廠可以增加產能。

Ⓓ 若中國工廠可以增快啟動時間。

解答

1. C 2. D 3. C 4. C 5. B

解題

1. 在第一封信件中提到 Both factories run "around the clock"six days a week. "Around the clock"是 24 小時的意思，因此答案為 C。

2. 在第一封信件中說明了宿霧工廠的強項為語言能力，而成都工廠的強項為再啟時間，因此在第二封回函裡 Neil 表示需再與兩家工廠談判，所以答案為 D。

3. 在第一封信件中說明到城都工廠的重啟時間為 8 小時，而宿霧工廠的重啟時間為 24 小時，因此 答案為 C。

4. 在第二封回信的最後 Neil 提到第一優先製造的談判，因此解答為 C。

5. Neil 只有在郵電中提及重要的因素但並未提及絕對因素，所以解答為 B。

1.**Capacity** (n) 產量，生產力

The productive capacity for this factory is 2000 unit per day.

這間工廠的產量為每天 2000 台。

2. **Around the clock** 繞鐘一圈，即為 24 小時的意思

Most fast food restaurants run around the clock.

大部份的速食餐廳營業 24 小時。

3. **Significantly** (adv) 重要地，重大地，意義重大地

The release of this prisoner has a significantly important meaning to the country.

釋放這個罪犯對這個國家來說是有非常重大的意義。

4.**Barrier** (n) 障礙，關卡

Lack of confidence is the biggest barrier for Lucy.

對 Lucy 來說，缺乏自信是她最大的關卡。

5. **Emphasis** (n) 重點，重要性

This interview places a great emphasis on communication skill.

這個面試非常重視溝通能力。

SAMPLE 7 Invoice

A Invoice

Bretec Water Filter Inc.		Date:	12/4/11
2000 North Second Street		Invoice Number:	AG77736211
Philadelphia,PA19090		Company:	Groupon International
		Project Code:	DC002
Groupon International			
1832 Six Street,			
Philadelphia, PA19090			
We would like to remind you that we have contractually agreed on payment of the amounts stated below by the end of this month.			
Product / Service Cost	Unit Price (USD)	Units	Subtotal(USD)
Water Filter	79.99	80	6,399.20
Installation	200	1	200.00
Pre-VAT Total			6,599.20
VAT 10%			659.92
Total Amount Due			7,259.12
Thank you for supporting Bretec Filter for all these years. Your prompt payment is appreciated.			

B Question about invoice

December 19th, 2011
Bretec Water Filter Inc.
2000 North Second Street
Philadelphia,PA19090

To whom it may concern,

Thank you for sending over the invoice. However, there are a few places that I don't understand and require for answers before paying the invoice. First of all, based on our contract agreement back in January 2007, the unit price for the filter should be set at $60. I believe you have used the regular price on the invoice. On the same contract, it also **stated** that my company could get an extra 5% off the Pre-VAT total. This invoice not only used the regular selling price for the filter, but also forgot to give us the discount we deserve. As a long-term customer, I do not appreciate being **looked upon to** pay higher costs which I am not required to do. I believe that it must be careless mistakes made by your new staff who is not familiar with the corporate discount contract. Please revise the invoice and send it to us again. I will pay the amount as soon as we receive the correct invoice.

Sincerely,
Rex Adams
Groupon International

Answer the Following Questions

1. Which of the following best described the situation?

A Bretex did not follow the contract.

B Groupon is satisfied with Bretex service.

C Groupon most likely will continue to work with Bretex.

D Bretex has increased the selling price for water filter.

2. According to the contract, how much should the pre-VTA total should be?

A $6,599.20

B $60

C $5,000

D $5,250

3. Which of the following is NOT true?

A Bretex's standard selling price for water filter is $79.99.

B Bretex does water filter installation for free.

C Rex is requesting Bretex to revise the invoice.

D Groupon had signed an agreement with Bretex to lock the selling price.

4. Which action will Bretex most likely NOT do?

A Revise the invoice.

B Discontinue the contract with Groupon.

C Check the contract.

D Send an appology letter to Groupon.

5.What action will Rex do next?

A Pay the amount as listed on the invoice.

B Request for a refund.

C Terminate the contract with Bretex.

D Wait for Bretex's response.

Answer

1. A 2. C 3. B 4. B 5. D

例文七 帳單發票

A 帳單發票

Bretec 淨水器公司		日期：	12/4/11
2000 North Second Street		發票號碼：	AG77736211
Philadelphia,PA19090		公司名稱：	Groupon 國際
		專案代號：	DC002
Groupon 國際			
1832 Six Street,			
Philadelphia, PA19090			
我們想要提醒您，依照我們之間的協定，請在本月底前支付下述的款項。			
產品 / 服務	單價（美元）	數量	小計（美元）
淨水器	79.99	80	6,399.20
組裝費	200	1	200.00
稅前總金額			6,599.20
10% 附加價值稅			659.92
支付總額			7,259.12
感謝您常年支持 Bretec 濾水器公司。並感謝您迅速的支付款項。			

B 對於帳單發票的疑問

2011 年 12 月 19 日
Bretec 淨水器公司
2000 North Second Street
Philadelphia,PA19090

敬啟者，

謝謝你送來發票。不過，有一些我不明白的地方，需要支付發票之前得到您的答案。首先，依據我們早在 2007 年 1 月的合同協議，淨水器的單價應定為 60 美元。我相信你在發票上已經使用了一般正常價格。在相同的合同裡也表示，我公司可以得到附加價值稅前小計的額外 5%折扣。此發票不僅使用了淨水器的一般單價，也忘了給我們應得的折扣。作為一個長期的客戶，我喜歡被期待我們須要支付更高的成本。我相信這應該是由您公司不熟悉企業折扣合同的新進員工所犯的粗心錯誤。請修改發票並再次發送給我們。一但我們收到正確的發票，我將支付金額。

真誠的，
Rex Adams
Groupon 國際

例題中譯

1. 下列何者最能敘述當時的情況？

A Bretex 未遵守合約。

B Groupon 很滿意 Bretex 的服務。

C Groupon 應該會持續與 Bretex 合作。

D Bretex 已調漲淨水器的價錢。

2. 根據合約，淨水器的稅前總價為何？

A $6,599.20

B $60

C $5,000

D $5,250

3. 下列何者不是事實？

A Bretex 淨水器的一般售價為 $79.99。

B Bretex 提供免費安裝服務。

C Rex 要求 Bretex 修改發票。

D Groupon 已與 Bretex 簽約已鎖定價格 .

4. Bretex 將不會採取下列哪個行動？

A 修改發票。

B 停止與 Groupon 的合約。

C 確認合約內容。

D 寄道歉信函給 Groupon。

5. Rex 將採取甚麼行動？

Ⓐ 支付發票上的金額。

Ⓑ 要求退款。

Ⓒ 終止與 Bretex 的和約。

Ⓓ 等待 Bretex 的回覆。

解答

1. A　2. C　3. B　4. B　5. D

解題

1. 從第二封的郵件裡我們可以瞭解到 Groupon 對 Bretex 的不滿，因此答案為A。
2. 第二封的郵件裡有提到合約裡的濾水器單價為 60 元，購買總數為 80 個，所以總價為 4800 元加 200 元組裝費，因此答案為C。
3. 在發票及郵件中均未提到免費安裝，因此答案為B。
4. Bretex 為賣方，所採取的行動應為確認合約、道歉及改發票，因此解答為B。
5. Rex 在回覆的郵件裡提到將在收到正確的發票後支付金額，因此解答為D。

字彙

1. **Invoice** (n) 發票，發貨清單

   List of product, unit price and total amount are required to be listed on the invoice.

   發票上規定需要列出貨物清單、單價及總價。

2. **Contractually** (adv) 受合同約束的

   The seller is contractually responsible for all products shipped out from his company.

   賣家依合同約束，對所有從公司寄出的商品均有責任。

3. **Prompt** (adj) 立即的，迅速的，準時的

   He is always prompt in responding to customers.

   他一向立即回覆客戶。

4. **State** (v) 說明，陳述

   Please state whether you are married or single.

   請說明你是已婚或是單身。

5. **Look upon** 期待

   I look upon education as an investment in the future.

   我期待教育是對未來的一種投資。

SAMPLE 8 Acquisition Process

A Follow up meeting request

From: Daniel Washinton <dwashinton@spaceklab.com>
To: Ariel Hoffman <ahoffman@northernsolution.com>
Date: March 17th, 2009

Dear Ariel,

It was my pleasure meeting with you yesterday. Our discussion over the future of Spaceklab was very **productive** and cleared up many financial concerns I had about our pending merger. I am very proud that Northern Solution sees the **potential** of Spaceklab and agreed to the merger. I have broken the news about Northern Solution's acquisition of Spaceklab to all staff. Although some concerns about the employee standing were raised, everyone is excited about the merger and look forward to working as a Northern Solution Employee.

Next week, I will be traveling around all week to meet with clients to let them know the future of Spaceklab, but if possible, I'd like to squeeze in another meeting with you to discuss other aspects of the merger including employee standing and benefits. I will be in contact after my travel schedule is firm, but please feel free to contact me anytime if you have any pressing issues.

I look forward to meeting you again.

Sincerely,
Daniel Washinton
Director
Spaceklab

B Response

From: Ariel Hoffman <ahoffman@northernsolution.com>
To: Daniel Washinton <dwashinton@spaceklab.com>
Date: March 17th, 2009

Dear Daniel,

Thank you for your email. Our lunch meeting yesterday was indeed **constructive** in laying out some **fundamental** guidelines for our merger. I was happy to have the chance to meet you personally.

I am pleased to hear that staff in Spaceklab is excited to hear about the merger, and we will be meeting again shortly, since we do have more to discuss in regard to Northern Solution **acquisition** policies and the welfare of your current staff. It is normal that staff will have the concern about the employee standing. Many employees are afraid if their position will still be available, if they will be laid off, or if the employee benefits will change. I can guarantee you that by becoming employees of Northern Solution, they will get more benefits than they have ever asked for.

My schedule is quite tight as well, but I have an open this coming Wednesday between the hours of 3 and 5, so if you are available, I think it would be wise for us to meet before your cliet meetings. I will be awaiting your call sometime soon.

Sincerely,
Ariel Hoffman
VP
Northern Solution

Answer the Following Questions

1. What was the purpose of the meeting?

A To discuss about potential business opportunities.

B To discuss about employee benefits.

C To discuss about the merger in between two companies.

D To solve the financial issues within Spaceklab.

2. What concerns Spaceklab's employee?

A The change of health insurance.

B Northern Solution might cut people.

C The work location might change.

D The workload might increase.

3. Which of the following is true?

A Northern Solution will be bought by Spaceklab.

B Northern Solution and Spaceklab will become joint venture.

C Spaceklab will change its company name to Northern Solution.

D After the merger, the company name will stay as Spaceklab.

4. What will be the purpose for the next meeting?

A Discuss about how to announce the merger.

B Regular weekly meeting.

C Discuss about employee standing and benefits.

D Reorganize the company.

5.Where will the next meeting be taken place?

- A Next Wednesday
- B Nothern Solution
- C Spaceklab
- D Not yet decided

Answer

1. C　　2. B　　3. C　　4. C　　5. D

例文八　併購處理

A 跟進會議要求

來自：Daniel Washinton<dwashinton@spaceklab.com>
此致：Ariel Hoffman<ahoffman@northernsolution.com>
日期：2009 年 3 月 17 日

親愛的 Ariel，

昨天我很高興會見您。我們對於 Spaceklab 未來的討論是非常有成效的，這消除了我對我們即將到來的合併案中財政上的不安。我非常自豪 Northern Solution 看到了 Spaceklab 的潛力並同意合併。我已經向 Spaceklab 全體員工發表了關於 Northern Solution 公司收購 Spaceklab 的訊息。雖然有人提出對於員工身分保障的擔憂，但每個人都對於合併表示興奮，並期待著成為 Northern Solution 的員工。

下週，我將前往各地與客戶會面，讓他們知道 Spaceklab 的未來。但是如果可能的話，我想與你安排另一次會議來討論其他合併案中其他的問題，包括員工身分的保障和利益。我會在確認行程後再度與您接觸。但若有緊急的事情，請隨時與我聯繫。

我期待著再次見面。

真誠的，

Daniel Washinton

經理

Spaceklab

B 回覆

來自： Ariel Hoffman <ahoffman@northernsolution.com>

此致 : Daniel Washinton <dwashinton@spaceklab.com>

日期 : 2009 年 3 月 17 日

親愛的 Daniel，

謝謝您的電子郵件。我們昨天的午餐會議，鋪設了一些基本準則，是個具有建設性的會議。我很高興有機會見到您本人。

我很開心能聽到 Spaceklab 的工作人員對於合併案表現出高興的態度，也很開心我們將很快地再次舉行會議，因為我們的確有必要對於 Northern Solution 收購政策和現存員工福利進行進一步的討論。既存員工對於自己的工作地位問題存有疑問是很正常的。許多員工會害怕合併後他們的職位是否還存在，他們是否會被解雇，或者員工福利是否會改變。我可以向您保證，成為 Northern Solution 的員工，他們將會得到從未有過的福利。

我的日程安排相當緊，但是本週三的 3 點到 5 點有空。因此，如果您也有時間，我認為這將是在您與客戶開會前，我們先行會面的明智時間。我將等待您的來電。

真誠的，
Ariel Hoffman
副總
Northern Solution

例題中譯

1. 本次會議的目的為何？

A 討論潛在的商業機會。

B 討論員工福利。

C 討論兩家公司之間的合併。

D 解決 Spaceklab 的財務問題。

2. Spaceklab 的員工關注的是甚麼？

A 醫療保險的變化。

B Northern Solution 可能裁員。

C 工作地點可能改變。

D 工作量可能會增加。

3. 下列何者為真？

A Northern Solution 將被 Spaceklab 併購。

B Northern Solution 和 Spaceklab 將成為合資公司。

C Spaceklab 將會更名為 Northern Solution。

D 合併後，公司名稱會維持叫 Spaceklab。

4. 下一次的會議主旨為何？

A 討論如何宣布合併。

B 每週定期會議。

C 討論員工定位及利益。

D 重組公司。

5. 下一次的會議將會在哪裡舉行？

A 下週三。

B Nothern Solution。

C Spaceklab。

D 尚未決定。

解答

1. C　2. B　3. C　4. C　5. D

解題

1. 從第一封郵件的開頭便提到此次會議的目的為討論合併案，因此答案為C。

2. 在第一封郵件裡提到 Spaceklab 的員工對於員工身分保障的擔憂，因此答案為B。

3. 由於 Spaceklab 是被 Northern Solution 併購，因此併購後應更名 Northern Solution，所以答案為C。

4. 第二封郵件裡提及下次的會議將會針對收購的政策與員工福利部份做討論，因此解答為C。

5. 在第二封郵件的最後有提到下次會議的時間，但並未提到會議的地點，因此解答為D。

A 字彙

1. **Productive** (adj) 多產的，富有成果的
   Most employees agree that it is very hard to have a productive meeting after 5pm.
   大部份的員工同意過了下午五點之後很難有富有成果的會議。

2. **Potential** (n) 淺力，淺能，可能性
   Human Resource Department needs to have the ability to see people's potential.
   人資部需要有能看見一個人潛力的能力。

3. **Constructive** (adj) 建設性的，積極的
   Interns should have a constructive attitude in order to get the chance to become official employees.
   實習生需要有積極的態度才有可能成為正式員工。

4. **Fundamental** (adj) 基本的，基礎的
   There is a fundamental difference in between these two theories.
   這兩個理論有著基礎上的不同。

5. **Acquisition** (n) 獲得，取得
   Knowledge is an invisible but valuable acquisition to everyone that gets education.
   知識，對受教者來說是一個不可見卻寶貴的收穫。

A Layoff Notification

To: Mr. Angus Lai
321 Hillsboro Dr.
Midland, TX 75236

Dear Mr. Lai,

We regret to inform you that Khuth Storage is forced to lay off 25% of its workforce. The decision was not an easy one, and it is one we highly regret. We value every employee and feel that the only fair way to do this is to **dismiss** the employees with the least amount of seniority. Unfortunately, you are among the last 25% of employees hired and; therefore, your employment with the company will be **terminated** as of April 1, 2012.

The current economic conditions have resulted in a severe lack of sales for our organization. Over the last 2 years, the number of new businesses created has declined sharply. While we are highly committed to our employees, we feel that the only way to **sustain** the business through this downturn is to cut expenses across the board.

Over the next 3 weeks, you will have the opportunity to meet with employment counselors who will hold meetings within the Human Resource Department. You will receive 4 weeks of **severance** pay at your normal weekly salary. To receive your severance, you will need to sign and return the release of claims documents in your packet. You will receive a package with this letter detailing the amount of benefits that you have as well as other helpful information.

We ask that you turn in your identification badge and any other company property on your last day of employment. Please be sure that we have your updated contact information on file. Again, we regret being forced into such a tough business decision. If you have any questions, you may schedule an appointment with me by sending an email to jkhuth@khuthstorage.com.

Sincerely,
Joanne Khuth
HR Director
Khuth Storage

B Challenge of Layoff

Dear Ms. Khuth,

I am writing you this email to ask for a **reconsideration** of the layoff. I am sorry to hear that the company was effected by the economy situation, and have no choice but to cut a certain percentage of the workforce. However, I do not think it is fair to let people go based on the seniority. For the three years I have worked with Khuth Storage, I have won the Employee of The Month for 5 times, and have never been late to work or leave work early due to personal reasons. I believe my performance is far better than several senior workers and my input to the company is undoubtable.

Furthermore, I also gave good suggestions to the company by cutting the cost and increasing the business to improve the efficiency. I not only see this work as a job, but also an opportunity to challenge myself and improve my personal skills. I hope I get the chance to grow with the company. Please reconsider the layoff and hopefully give the hard workers a chance to stay with the company. If possible, please allow me to arrange a personal meeting with you. I hope I can get a chance to convince you in person.

Best regards,
Angus Lai

Answer the Following Questions

1. What is the main point of the first email?

A The economy condition forces Khuth Storage to shrink their business.

B Khuth Storage will cut 25% of its workforce.

C Khuth Storage plans to lay off Angus.

D Severance package will be given to all employees that got let go.

2. Which of the following will Joanna most likely do?

A Keep hiring Angus.

B Increase the severance package for Angus only.

C Find Angus another job.

D Arrange a private meeting with Angus.

3. Which word best describes the mood of 2nd email?

A Disappointing

B Hypocritical

C Aggressive

D Encouraging

4. Which of the following is NOT true?

A Angus thinks it is unfair to cut the workforce by seniority.

B Business has been bad for Khuth Storage for the past 2 years.

C There is no possibility that Angus can get his job back.

D All employees who get laid off can get one month worth of salary.

5. According to Angus, which of the following quality doesn't Angus have?

A Challenging

B Cheerful

C Fearless

D Creative

Answer

1. C 2. D 3. C 4. C 5. B

例文九 裁員及再次機會挑戰

A 裁員通知

此致：Angus Lai 先生
321 Hillsboro Dr.
Midland, TX 75236

親愛的 Lai 先生，

我們很遺憾地通知您，Khuth Storage 被迫裁員其員工總數的 25%。這是個不容易的決定，我們也非常遺憾。我們重視每一位員工，並認為唯一公平的解雇方式是以資歷來作判斷。很不幸的，您的資歷是落在最後 25% 之內，因此，您的工作將在 2012 年 4 月 1 日終止。

目前的經濟狀況已經導致了我們銷售上嚴重的缺乏不足。過去 2 年，新業務的數量急遽的下降。雖然我們對我們的員工有高度的承諾，但為了能夠度過這次的經濟衰退並維持業務的唯一辦法就是統一削減開支。

在接下來的 3 週，您將有機會與人力資源部所舉辦的會議中與就業輔導員會談。您將享有 4 週的解僱費，並在您正常的週薪時支付。為收到您的遣散費，您需要回簽在文件裡的放棄訴訟請求書。您將收到一個包裹，包裹將詳細說明您的補助金額以及其他有用的信息。

我們要求您在最後一天交出您的身份徽章和任何其他公司的財產。請確認我們有您更新過的聯繫資料。我們再次對這樣艱難的商業決定感到遺憾。如果您有任何問題，您可以藉由發送電子郵件到 jkhuth@khuthstorage.com 與我安排會面。

真誠的，

Joanne Khuth

人力資源總監

Khuth Storage

B 挑戰裁員

尊敬的 Khuth 女士，

我寫這封電子郵件希望您可以重新考慮裁員的決定。我很遺憾聽到這家公司由於經濟形勢的影響，必須做出一定比例的裁員。不過，我不認為以資歷長短作為裁員的標準是公平的。我在 Khuth Storage 工作的三年以來，已經贏得了 5 次每月的最佳員工，從來沒有遲到或因個人因素早退。我相信我的表現遠比幾個資深員工更好，我對公司的貢獻也是毋庸置疑的。

我同時也給了公司很好的建議，使公司提高了效率也降低成本，進而提高企業的經營。我不僅僅認為這是一份工作，但也視為是個挑戰自己的機會，提高自己的個人技巧。我希望我有機會與公司共同成長。請重新考慮裁員和給辛勤工作的職員有留在公司的機會。如果可能的話，請允許我安排與您面談。我希望我能有機會親自來說服您。

最好的問候，

Angues Lai

例題中譯

1. 第一篇電郵的主旨為何？

A 經濟狀況迫使 Khuth Storage 縮小他們的生意。

B Khuth Storage 將裁員 25%。

C Khuth Storage 將裁員 Angus。

D 遣散費將被提供給所有被裁員的員工。

2. Joanna 最有可能採取甚麼行動？

A 續任 Angus。

B 增加 Angus 的遣散費。

C 替 Angus 找另一份工作。

D 安排與 Angus 的私人會議。

3. 下列何字最能闡述第二封電郵的情緒？

A 失望

B 虛偽

C 積極

D 鼓舞人心

4. 以下何者為非？

A Angus 認為以年資最為裁員標準是不公平的。

B Khuth Storage 過去兩年的生意都不好。

C Angus 不可能復任之前的工作。

D 所有被裁員的員工皆可領一個月的遣散費。

5. 根據 Angus，他沒有什麼特質？

A 挑戰

B 快樂

C 不畏懼

D 創意

解答

1. C　2. D　3. C　4. C　5. B

解題

1. 解答的 A B D 皆為第一封電郵裡向 Angus 解釋需要解聘他的原因，但其主旨為解聘 Angus，因此答案為C。

2. 在第一封電郵裡 Joanna 有提到任何人有問題皆可以提出與她私人面談的要求，這也正是第二封電郵 Angus 的要求，因此答案為D。

3. 在第二封電郵中 Angus 很積極地說明 Joanna 不應該裁員他的原因，並要求私人會面以改變 Joanna 的想法，因此解答為C。

4. 在兩篇電郵中都未提到不可能復職，因此解答為C。

5. Angus 不畏懼地提出了第二封電郵來挑戰裁員的結果，並在文中提到他具有創意的提案以及挑戰自己的特質，但並未在文裡提到“快樂”的部份，因此解答為B。

A 字彙

1. **Dismiss** (v) 解雇，開除

   David was dismissed because of his poor performance.

   David 因為他不好的表現被解雇。

2. **Terminate** (v) 使…終止，使…結束

   The contract cannot be terminated unless you pay the penalty.

   除非你付罰金，否則契約是不能被終止的。

3. **Sustain** (v) 維持，保持

   We need every visitors' cooperation to sustain the condition of the historical site.

   我們需要每位遊客的合作以維持古蹟的狀況。

4. **Severance** (n) 遣散費

   It is illegal to dismiss an employee without paying severance.

   解雇員工但不付遣散費是不合法的。

5. **Reconsideration** (n) 重新考慮

   With reconsideration, the couple decided to give their marriage another chance.

   重新考慮後，這對夫妻決定再給他們的婚姻一個機會。

SAMPLE 10 Company Event

A Company Event Invitation

To all staff,

The company will be holding the annual sales meeting in Las Vegas, NV from Dec. 16th to 18th. All staffs are invited. The company will pay for all transportation, including flights to Las Vegas and buses in Las Vegas. For **domestic** travelers, please file a copy of your driver's license to HR by Oct. 30th. For international travelers, please file in a copy of your passport with a **valid** date for at least 6 months from Dec. 18th to HR by the end of this week. If your wish to take family members with you, it is allowed but will be charged for travel fees. For detailed information, please contact your HR. Note that due to the process time, if documents are not filed by requested dates, there might be a possibility that you will miss the chance to join the event.

Hotel **accommodation** has been booked at the Caesars Palace. Senior managers will have their own rooms. All other staffs will be sharing twins. Due to the large number of staffs, we cannot take requests for your roommate. It will be **the luck of the draw**. If you plan to pay extra and travel with your family, you will be **guaranteed** to share rooms with your family.

We're planning to return transport on Dec. 18th. All staffs will be traveling to the airport by bus together at 10:00 am. If you have a late evening flight and prefer to go to the airport by yourself, please inform us while you file the documents to HR.

Best regards,
Alexie Yang
ayang@wwscorp.com

B Response and detail confirmation

Dear Alexie,

I hope you haven't booked the flight tickets yet because I'm planning to go to the event with my family by car. No flight tickets will be needed. However, since I will be taking my wife and my 2 kids, I wonder if it's possible to arrange thru HR as well. If not, I wonder if I can book the hotel by myself and still join the company events. The hotel might not be at Caesar's Palace but somewhere near by.

I will turn in the documents by the end of the day. Thank you.

Best regards,
Joseph Lee
jlee@wwscorp.com

Answer the Following Questions

1. Which department does Alexie most likely work in?

Ⓐ Sales & Marketing

Ⓑ Research & Development

Ⓒ Production

Ⓓ Human Resource

2. What is the difference in between international and domestic travelers?

Ⓐ Meeting agenda.

Ⓑ Transportation after arriving Las Vegas.

Ⓒ Documents that need to be filed to HR.

Ⓓ Whether can bring family members or not.

3. What is Joseph asking for?

Ⓐ Extra flight tickets for his family.

Ⓑ Special hotel arrangement.

Ⓒ A rental car to drive to Las Vegas.

Ⓓ Non of the above.

4. When do international travelers need to turn in the paper works?

Ⓐ October 30th

Ⓑ December 16th

Ⓒ June 16th

Ⓓ End of the week after receive the announcement.

5.Which of the following is true?

(A) All staffs are required to go to the airport together on December 18th unless report to HR in advance.

(B) The purpose of this trip is to appreciate employees' hard work for the year.

(C) Joseph will not bring his family if the company doesn't agree to let him book his own hotel.

(D) All employees are required to bring their family members.

Answer

1. (D) 2. (C) 3. (B) 4. (D) 5. (A)

例文十 公司活動

A 公司活動邀請

所有的工作人員，

該公司將在 12 月 16 日至 18 日於內華達州拉斯維加斯舉行年度銷售會議。全體員工皆受邀參加。該公司將支付所有的運輸費用，包括飛往拉斯維加斯的機票及拉斯維加斯城內的巴士。若您是國內的旅客，請在 10 月 30 日前提交駕駛執照的副本給人力資源部。若您是國際旅客，請在本週末結束前繳交從 12 月 18 日算起尚有 6 個月有效期限的護照影本給人力資源部。如果您想帶家人和你在一起，這是允許的，但將收取旅行費用。有關詳細信息，請聯繫您的人力資源部。請注意，由於處理時間上的關係，如果未在要求提交日期前繳交文件，您將可能會錯過參加此次活動的機會。

酒店已經預訂在凱撒宮。高級管理人員將擁有自己的房間。所有其他員工將分享雙人房。由於工作人員數量龐大，我們無法接受指定室友的請求，這將會以抽籤方式決定。如果您打算支付額外費用並與您的家人同行，您將保證與家人分享房間。

我們的回程運輸計劃是在 12 月 18 日。所有的工作人員將一同於早上 10:00 乘坐巴士前往機場。如果您是晚上的班機並傾向自己去機場，請在繳交文件給人力資源部時一併告知我們。

最好的問候，

Alexie Yang

ayang@wwscorp.com

B 回覆及詳細內容確認

親愛的 Alexie，

希望您還沒有預訂機票。我打算和家人開車去參加這次的活動，所以沒有機票的需求。然而，我會帶我的妻子和我的兩個孩子。我不知道是否可能請人力資源部直接安排酒店。如果不行，我是否可以自己預訂附近的酒店，並仍然參加公司的活動。酒店可能不會在凱薩宮，而是附近的某個飯店。

我將會在今天結束前繳交文件。謝謝。

最好的問候，

Joseph Lee

jlee@wwscorp.com

例題中譯

1. Alexie 可能是在下列哪一個部門工作？

A 市場行銷

B 研究與開發

C 生產

D 人事

2. 何者為國際與本地遊客的不同？

A 會議議程

B 抵達拉斯維加斯後的交通

C 需要提交給人事部的資料

D 是否可帶家人

3. Joseph 的要求是甚麼？

A 給他家人的額外機票

B 特殊的酒店安排

C 開去拉斯維加斯的租賃車

D 以上皆非

4. 國際遊客需要在何時繳交書面資料？

A 10 月 30 日

B 12 月 16 日

C 6 月 16 日

D 收到公告的週末

5. 以下何者為真？

(A) 除非有事先與人事報備，否則所以遊客皆須在 18 號一起前往機場。

(B) 這趟旅遊的主旨是要慰勞員工的辛勞。

(C) 若公司不讓 Joseph 自己訂酒店，他將不會帶家人。

(D) 所有員工皆被要求帶家人。

解答

1. (D) 2. (C) 3. (B) 4. (D) 5. (A)

解題

1. 由於 Alexie 是主辦此次活動的單位，因此最有可能的服務單位為人事部。答案為(D)。

2. 在第一封電郵裡提到國際旅客需要提繳護照影本而本地旅客需要提繳駕照影本，其它部份皆相同，因此答案為(C)。

3. 在第二封電郵中，Joseph 提到希望帶家人一起去，因此需要特殊的酒店安排。解答為(B)。

4. 在第一封電郵中提到國際旅客需在週末結束前繳交從 12 月 18 日算起尚有 6 個月有效期限的護照影本給人力資源部，因此解答為(D)。

5. 在第一封郵件的最後提到所有的工作人員將一同於早上 10:00 乘坐巴士前往機場 ，因此解答為(A)。

A 字彙

1. **Domestic** (adj) 國內的，本土的
   Every citizen should support domestic agriculture by purchasing local produce.
   所有市民都應該購買當地的農產品以支持國內農業。

2. **Valid** (adj) 正當的，有效的
   A valid student ID is required when purchasing the discount movie ticket.
   當購買折扣電影票時，有效的學生證是必要的。

3. **Accommodation** (n) 住處，住房
   Accommodation arrangement is often a big challenge for travelers.
   住房安排通常對旅客來說是一大挑戰。

4. **The luck of the draw** 靠運氣的
   You can't control the weather situation during your honeymoon. It is just the luck of draw.
   你不能掌控你度蜜月時的天氣。這是靠運氣的。

5. **Guarantee** (v) 保證，擔保
   The online store has guaranteed that all products can be delivered within 24 hours.
   線上商城保證所有貨品皆可以在 24 小時內到貨。

A Change of Credit Card Agreement

Date: June 23, 2013

CHANGE TO YOUR CREDIT CARD AGREEMENT

Please read this document carefully and keep it with your Credit Card Agreement. Except as **amended** below, the terms of our Credit Card Agreement remain in effect. If there is a **conflict**, the terms in this Amendment will **pre-vail**.

TRANSACTION FEE FINANCE CHARGES

We are increasing certain **transaction** fees on your account.

Amendment to Your Credit Card Agreement:

Effective August 1, 2013, the transaction fee we assess on each of the transactions identified below will be equal to 2% of each such transaction.

- ATM Cash Advances
- Balance Transfers
- Bank Cash Advances
- Check Cash Advances
- Wire Transfer Purchases

CHANGES TO YOUR WORLDPOINTS REWARDS PROGRAM

Effective August 1 2013, the following changes will be made to your account:

Travel Rewards

You will no longer be able to redeem Points for free hotel rewards. Instead, you can now use our more popular AnytimeAir Rewards. AnytimeAir Rewards allows you to pick the flight of your choice at any time you wish to travel with no restrictions.

B Cancel Account

Date: July 6, 2013
Customer Service
Bank of The Galaxy
16600 E Palisades Blvd
Fountain Hills AZ 85268
Re: Closing account on Frequent Traveler Visa Account Number 3317 0714 0023 2209

Dear Sir or Madam:

I have received the Change of Credit Card Agreement at the end of last month. Since I can no longer benefit from the original rewards that I applied the credit card for, I am sending you this letter as my official notice that I will be closing my account 3317 0714 0023 2209 by the end of the month with Bank of the Galaxy.

Account balance was paid on date and was confirmed that this payment was received on the posting date. I also am enclosing my destroyed credit card.

To my knowledge, all my fiscal responsibilities with this credit card account have been fulfilled. Therefore, please close my account and include a **notation** in the report to the credit **bureaus** that the account was "closed by request of cardholder." Once this is done, please send me written confirmation of the closure of my account.

Last but not least, I have enjoyed using the Frequent Traveler VISA and satisfied with the service from Bank of the Galaxy for many years. However, since I am a frequent domestic traveler and free hotel rooms are important for me, I have no choice but to switch credit card company. Therefore, if there is any other credit cards from your bank that give out points which can redeem for domestic hotel rooms, please let me know. I very likely will apply again.
Thank you.

Sincerely,

Albert Anderson

Answer the Following Questions

1. What is the letter for?

A Air mile reward points statement.

B Credit card agremment change notice.

C Hotel reward program revision.

D Frequent traveler reward notice.

2. Which of the following will not be changed?

A Hotel reward points

B Air mile reward points

C Transaction fees

D Late payment penalty

3. What is Albert's request?

A Eliminate the changes and keep using the credit card.

B Cancel the current credit card and apply for a new one.

C File a complain to the bank but still continue to use the credit card.

D Cancel the current credit card and ask for another card that fit his needs.

4. When was the agreement received?

A July 6th

B End of June

C June 23rd

D End of July

5. What will happen next?

A Albert will continue to use Frequent Traveler VISA after negociate with the bank.

B The costs for using the Frequent Traveler VISA will increase.

C The bank will waive the transaction cost for Albert to keep this client.

D The bank will give Albert extra AnytimeAir Rewards if Albert continue to use the credit card.

Answer

1. B 2. D 3. D 4. B 5. B

例文十一 銀行相關

A 信用卡協議更改

日期：2013 年 6 月 23 日

信用卡協議更改

請仔細閱讀本文件，並與你的信用卡協議一同保留。除以下修訂，我們的信用卡協議的條款仍然有效。如果有衝突，則以這個修訂的條款為準。

交易費用財務費用

我們將增加您帳戶的一些交易費用。

修訂的信用卡協議：

2013 年 8 月 1 日起生效，每個以下指定交易的交易費將為每次交易的 2%。

- ATM 預借現金
- 餘額轉移
- 銀行現金墊款
- 支票現金墊款
- 電匯購買

WORLDPOINTS 獎勵計劃變更

以下對於您帳戶的更改將在 2013 年 8 月 1 日起生效：

旅行獎勵

您將不再能夠以積分兌換免費酒店的獎勵。相反，你現在可以使用我們的更受歡迎 AnytimeAir 獎勵。 AnytimeAir 獎勵讓你挑選你所選擇的機票，且將沒有任何航班或時間的限制。

B 取消戶頭

日期：2013 年 7 月 6 日
顧客服務處
銀河銀行
16600 E Palisades Blvd
Fountain Hills AZ 85268

Re：取消旅遊常客ＶＩＳＡ信用卡帳號 3317 0714 0023 2209 帳戶

尊敬的先生或女士：

我在上個月末有收到信用卡協議的變更。由於原有獎勵不再受益，這封信將是我將在月底關閉我在銀河銀行帳號 3317 0714 0023 2209 的正式通知。

帳戶餘額已經支付並已確認在該付款日期入帳。我同時附上我剪毀的信用卡。

以我的理解，我與此信用卡帳戶的財政責任已經完成。因此請關閉我的帳戶，並在給信用局的報告上加註"由持卡人請求關閉"。

以上完成後，請給我關閉帳戶的書面確認書。

最後，多年來，我一直是旅遊常客 VISA 的用戶並享受及滿意來自銀河銀行的服務。但是，由於我經常國內旅行，免費酒店房間對我很重要，我也只好改用信用卡公司。因此，如果有任何其他由貴銀行發卡的信用卡可以兌換國內酒店房間點數，請讓我知道。我很可能會再次申請。

謝謝。

真誠的，

Albert Anderson

例題中譯

1. 這封信的用途是甚麼？

A 航空里程表。

B 信用卡條約更改通知。

C 酒店獎勵計畫更改。

D 旅遊常客獎勵通知。

2. 下列何者不會更改？

A 酒店獎勵點數

B 航空里程點數

C 手續費

D 滯納金

3. Albert 的要求為何？

A 取消更改，並繼續使用此信用卡。

B 取消此信用卡，並申請一張新的信用卡。

C 提交一份投訴到銀行，但仍繼續使用的信用卡。

D 取消當前的信用卡，並要求適合他需要的另一張卡。

4. 這份同意書是何時收到的？

A 7 月 6 號

B 6 月底

C 6 月 23 號

D 7 月底

5. 接下來會發生什麼事？

A 艾伯特將會再與銀行協商後繼續使用旅遊常客 VISA。

B 使用旅遊常客 VISA 的成本將增加。

C 為了保留 Albert 這個客戶，銀行將免收交易手續費。

D 如果 Albert 持續使用這張信用卡，銀行將提供額外的 AnytimeAir 獎勵。

解答

1. B　2. D　3. D　4. B　5. B

解題

1. 第一篇的標題即為信用卡協議更改，因此答案為B。
2. 在信用卡更改協議中沒有提到滯納金，因此不會被更改。答案為D。
3. 在第二封電郵中，Albert 要求取消信用卡，並在最後一個段落提到若有可以對換國內酒店點數的信用卡則會再提出申請，因此解答為D。
4. 第一封信件是在 6 月 23 日寄出，而從第二封電郵的開頭可得知這個協議是在 6 月底收到，因此答案為B。
5. 在第一封的協議中提到了各種手續費的增加，因此如果繼續使用此信用卡，則成本將會增加。解答為B。

字彙

1. **Amend** (v) 修改用詞，修正
   The country's constitution was amended to allow women to vote.
   國家的憲法被修正為女性可以投票。

2. **Conflict** (n) 衝突，抵觸
   The conflict between science and religion cannot be avoided.
   科學與宗教間的衝突是不可避免的。

3. **Prevail** (v) 勝過，佔上風
   Justice always prevails.
   正義永遠戰勝。

4. **Notation** (n) 記號，符號
   Memorizing basic musical notation is required for all elementary students.
   所有小學生都被要求暗記基本的音樂符號。

5. **Bureaus** (n) 政府的局處
   It is required to bring a photo ID to all government bureaus.
   現在到政府各局處都需要攜帶身分證。

SAMPLE 12 Business Plan

A New Business Model

To: Frank Clerk
From: Edward Chen
Date: March 9, 2015
Subject: E-commerce and retirement planning

Dear Frank,

After doing some research, I found that the retirement planning is the next big step for our company to invest in. Statistics shows that Taiwan now is **gradually** moving into aging population, and the low birth rate will shift the needs for senior care dramatically. Even so, the idea of retirement planning is not yet **popularized**. People still rely on the government pre-set retirement plan which is estimated to collapse within 20 years. There is a clear vision that we should start building the E-system that allows individuals to forecast their future living by inputting the assets they have in hand on the Internet by themselves, and get an idea of what kind of living they will be having if they invest their money into our retirement plan. Instead of using human resource, using network system gives each individual more freedom to create the living they want, and furthermore put more assets into our company. It can also lower human resource for regular accounts, and allow our banking specialists to focus on the big customers.

With all that said, this will not be an easy project. Developed countries, such as England and America have been working on this for years, and still many improvements can be done. For us, I think at least 5 years of **ground building** work will need to be done. However, I truly believe that is where the future lies. Please take considerations to start the project, and I will be honored to be the leader of it.

Sincerely,
Edward Chen
Product Development

B Kick Off Meeting Request

To: Edward Chen
From: Frank Clark
Date: March 10, 2015
Re: E-commerce and retirement planning

Dear Edward,

I am really glad you brought up this project. I have the same thought as you and believe that retirement plan development will be a big step for our company. Please put together a project development team by the end of next week and set up a **kick off** meeting. In the meeting, I request to see a draft plan that shows the estimated investment we need, including assets, human resources, timetable, and competition. I also want more detailed information that shows the estimation of **composition** of the population and the income and available assets for each household in Taiwan for the next 40 years.

I request all board members and also the following members to attend the meeting. Director of IT – James Patterson, Director of Human Resource – Emily Fong, Director of Sales – Catherine Chung. You can ask my assistant to help arrange the meeting accordingly.

We will decide who the project leader will be during the kick off meeting, and I believe you have the biggest chance to take onto the responsibility.

I look forward to having the meeting with you.

Regards,
Frank Clark
CEO

Answer the Following Questions

1. What is the email regarding to?
   - A Employee retirement plan
   - B Government pre-set retirement plan
   - C New product development
   - D Senior care

2. Why does Edward think this plan is a good investment for the company?
   - A The construction of the society is changing.
   - B The country has a very good pre-retirement plan.
   - C People prefer to have the freedom to control their own assets.
   - D This is a brand new idea that no country has thought about it before.

3. Which of the followings is NOT true?
   - A Frank is supporting the new business plan and wants to start it ASAP.
   - B Developed countries such as England and the Unite States have had this kind of products for years.
   - C Edward thinks it is necessary to build the plan under e-system so people have freedom to control their assets thru internet.
   - D Frank agrees to name Edward the project manager.

4. Who will NOT attend the meeting?

(A) Director of IT.

(B) Emily Fong

(C) Frank’s assitant

(D) Frank

5. What will Edward bring to the meeting?

(A) Competitors’ plans

(B) Attendees’ basic information

(C) A draft plan

(D) Available assets for the next 10 years.

Answer

1. (C) 2. (A) 3. (D) 4. (C) 5. (C)

例文十二 事業計劃

A 新商業模式

此至：Frank Clark

來自：Edward Chen

日期：2015 年 3 月 9 日

主題：電子商務和退休規劃

親愛的 Frank，

在我做一些研究之後發現，退休規劃是為我們公司接下來投資的一大步。統計顯示，台灣目前正逐步進入人口老齡化，低出生率使得老年護理將被大量需求。即便如此，退休規劃的理念尚未普及。人們仍然依靠政府預先設定的退休計畫，但政府的計畫估計將在 20 年內崩盤。這是一個明確的目標，我們應該開始建立 E- 系統，允許個人通過自己輸入他們手頭上的資產來預測其未來的生活，並預見藉由投資他們的錢到我們的退休計畫可以得到甚麼樣的生活。與其使用人力資源，通過網絡系統賦予每個個體更多的自由來創造他們想要的生活，並進一步把更多的資產注入本公司。這不但可以降低一般帳戶的人力資源，也可以讓我們的銀行專家專注於服務大客戶。

雖說如此，這不會是一個簡單的計畫。已開發國家如英國和美國已經在這方面涉獵多年，仍然有很多的改進的空間。對於我們來說，我認為至少有 5 年的基礎建設需要做。不過，我認真相信這正是未來的走向。請考慮採取啟動該計畫，我將榮幸地成為它的領導者。

真誠的，

Edward Chen

產品開發

B 要求啟動會議

此至：Edward Chen

來自：Frank Clark

日期：2015 年 3 月 10 日

回覆：電子商務和退休規劃

親愛的 Edward，

我真的很高興你提出了這個計畫。我和你有同樣的想法相信退休計畫的發展將是我們公司的一大步。請在下週結束前組成一個開發團隊，並成立啟動會議。在會議上，我要求看到草案書，內容包括預計的投資，包括資產、人力資源、時間表和競爭對手。我也希望可以看到接下來 40 年更詳細的台灣人口構成及每個家庭估計的收入和可用資產。

我要求所有董事會成員，及以下成員出席會議：ＩＴ總監 – James Patterson、人力資源總監 – Emily Fong、銷售總監 – Catherine Fong。你可以請我的助理幫你安排會議。

我們將在啟動會議時決定誰是該計畫負責人，但我相信你將最有機會接手此責任。

我期待著與你相見。

問候，

Frank Clark
首席執行官

例題中譯

1. 這份電郵是關於什麼？

A 員工退休計畫

B 政府預設退休計畫

C 新產品開發

D 老人照顧

2. 為什麼 Edward 認為這是一個對公司很好的投資計畫？

A 社會的結構正在改變。

B 國家有很好的預設退休計畫。

C 人民喜歡有控制自己資產的自由。

D 這是一個連其他國家都沒有的全新計畫。

3. 以下何者為非？

A Frank 支持這個計畫並希望可以儘早開始。

B 英國、美國等已開發國家已有類似計畫很多年了。

C Edward 認為此計畫應利用電子商務開發，使人民可以藉由網路自由控制資產。

D Frank 答應任命 Edward 為計畫負責人。

4. 誰將不會參加會議？

A IT 總監

B Emily Fong

C Frank 的助理

D Frank

5. Edward 將帶甚麼到會議上？

A 競爭者的計畫。

B 出席者的基本資料。

C 計畫草案。

D 未來 10 年的可用資產。

解答

1. C 2. A 3. D 4. C 5. C

解題

1. 在第一封電郵中 Edward 所提出的是以“退休計畫”為主軸的新產品，因此解答為 C 。

2. 在第一封的電郵中提到社會老齡化及少子化，皆為社會結構的改變，因此答案為 A 。

3. 在第二封 Frank 的回信的最後，Frank 提到將在啟動會議時決定計畫負責人，因此解答為 D 。

4. 在第二封回信中，Frank 列出了希望出席的人選，並讓 Edward 請 Frank 的助理 Emily 幫忙，但並未要求 Emily 參加會議，因此答案為 C 。

5. 在第二封回信中，Frank 要求 Edward 做一份計畫草案，內容需包括 A, B 與 D 因此解答為 C 。

A 字彙

1. **Gradually** (adv) 逐漸地

   Birth rate has been gradually decreased every year for the past 10 years.

   過去十年的出生率逐年遞減。

2. **Popularize** (v) 普及，推廣

   Internet was not popularized 15 years ago.

   15 年前網絡並不普及。

3. **Ground building** 基礎建設

   A solid ground building work is necessary for any job to become successful.

   任何工作成功的要件都需要堅固的基礎建設。

4. **Kick off** 開始，起始，啟動

   The president will personally host the kick off meeting for this new product.

   總裁將親自主持這個新產品的啟動會議。

5. **Composition** (n) 組成，構成

   A wrong chemical composition might cause serious damage to human body.

   錯誤的化學組合有可能對人體造成嚴重的傷害。

SAMPLE 13 Organization Introduction

A Animal Shelter Introduction

Purr Rescue is an all-volunteer non-profit animal rescue organization, **comprised** entirely of experienced volunteers who collectively have rescued hundreds of companion animals and placed them in loving homes.

United and inspired by our compassion for all animals, we focus our talents on saving unadopted cats and dogs destined to be **euthanized** in local shelters. Each rescued animal will be provided with veterinary attention, including examination, immunizations, and spay. Then the rescued dog or cat is placed into one of our safe haven foster homes where he receives basic family member training. Most of our dogs and cats are well-behaved and are ready for adoption.

We were founded in November 1998. Since forming, the group has rescued over 4400 dogs and cats, and has provided assistance to other rescue groups, shelters and owner's dogs and cats, to the tune of 4500 animals! A key part of our philosophy and practice is to build **alliances** with other cooperative rescue organizations, shelters, veterinarians, media, and caring companies. By working together, we can save more lives and eliminate the unnecessary deaths of so many adoptable animals while **enhancing** the lives of people as well.

If you are looking for a furry best friend, let us know! Tell us about yourself and the type of companion animal you would like to adopt. We have many wonderful dogs and cats available for adoption. We will find the perfect match or help you in your search with our extensive network of other rescue organizations.

B Ask to adpot a cat

To whom it may concern,

My family and I are looking to adopt a kitten to become one of our family members. Let me start by introducing my family. We are a family of 4, my wife Stacy, a 4 year-old son Andrew, a 2-year-old daughter Kim and myself. I am an electronic engineer at PG&E, and my wife is a stay home mom who commits all her time to this family.

3 months ago, we moved to Atlanta from Chicago due to my work relocation. We fell in love with this city and decide to settle down. We recently purchased our first home here and plan to spend our life time in this lovely city. My family and I are all animal lovers and have been wanting to have a furry family member for years. However, due to lack of personal property and time, we hesitated to adopt any kind of pets. Luckily, the situation has changed, and we think

it cannot be a better time for us to welcome a kitty to our family.

Growing up, both my wife and I were **surrounded** by dogs and cats. My parents have been foster parents for cats for over 20 years. Therefore, there will be no problem for me to take care of kittens even if they were not trained. I think I had learned how to be a caring and sensitive person because I was surrounded by rescued animals while growing up. And now, I am hoping that I can give my kids the same experience and furturemore make them a better person.

With all that said, please consider our request to adopt one of your little kittens. I guarantee that we will do our best to give this little one a family full of love. We are willing to accept any kind of kittens even the ones that have conditions. We look forward to hearing back from you soon.

Sincerely yours,
Jack Thompson

Answer the Following Questions

1. What kind of organization is Pur Recue?

A It provides perninent homes for animals.

B It gives medical care for pets for free.

C It helps cats and dogs to fine their new homes.

D It sells pets at a reasonable price.

2. What is NOT the purpose of this article?

A To make people understand how animal shelters work.

B To show people what the organization is capable of.

C To encourage people to adpot pets from them.

D To show the difficulty of running an animal rescue organization.

3. What does "foster home" mean?

A The permanent home for someone.

B The shelter.

C A house in the woods.

D A temperary home.

4. Why do the Thompsons want to adpot a cat?

A Because they do not have time to take care of a dog.

B Because kittens are more friendly than dogs.

C Because Mr. Thompson wants his kids to experience what he had experienced when he was a kid.

D Because they can put less attention to a cat than to a dog.

5. What most likely will happen next?

A The Thompson's will be interviewed by Pur Rescue.

B The Thompson's gets to choose the kitten they like.

C The Thomson's will change their mind and get a dog instead.

D The Thompson's will become volenteers for the organization.

Answer

1. C 2. D 3. D 4. C 5. A

例文十三 組織介紹

A 組織介紹

Pur 救援是一個全志願非營利性的動物救援組織。組織成員的經驗豐富，已經救出了數百隻寵物，並把他們安置在充滿愛的家園裡。

我們對動物的關愛讓我們深受鼓舞，團結一致並發揮我們的才能拯救未領養的貓狗免受安樂死於當地的收容所中。每個被救出的動物，會得到獸醫的關注，包括檢查、免疫和絕育。獲救的狗或貓將被安置於我們的寄養家庭。在那裡他將會接受基本的家庭成員培訓。我們的大多數狗和貓已有適當的行為，並已可被領養。

我們成立於 1998 年 11 月。成立以來，本組織已救出超過 4400 隻貓狗，並協助其他救援組織、庇護所和業主的貓狗等高達 4500 隻！

我們的理念和實踐中一個重要部分就是建立與其他合作救援組織、庇護所、獸醫、媒體和愛心企業的聯盟。通過共同努力，我們可以挽救更多的生命和消除許多收養動物不必要的死亡，同時提高人們的生活。

如果你正在尋找一個毛茸茸的好朋友，請讓我們知道！告訴我們關於你和你想找的寵物類型。我們有很多種類的貓狗可供收養。我們會替你找到最完美的匹配，或幫助您在我們廣泛的其他救援組織網絡中搜索。

B 詢問收養一隻貓

敬啟者，

我的家人和我都希望領養一隻小貓成為我們的家庭成員。首先，讓我由介紹我的家庭開始。我們一家四口，我的妻子 Stacy，4 歲的兒子 Andrew，2 歲的女兒 Kim 和我自己。我是一名 PG&E 的電子工程師，我的妻子是一個貢獻自己所有的時間給這個家庭的家庭主婦。

3 個月前，由於我工作的調派，我們從芝加哥搬到了亞特蘭大。我們愛上了這座城市，並決定定居下來。我們最近在此購買了我們的第一個家，並打算在這個美麗的城市過我們的生活。我的家人和我都是動物愛好者，並一直想有個毛茸茸的家庭成員。然而，由於缺乏個人房產和時間，我們無法領養寵物。幸運的是，情況發生了變化，我們認為現在是我們歡迎一隻貓咪成為我們家一份子最好的時間。

在長大的過程裡，我的妻子和我都被狗和貓所包圍。我的父母一直是貓的寄養父母，超過 20 年。因此，對我來說，即使是照顧沒有受過訓練小貓也不是問題。我想，成長在一個週圍都是被救援動物的環境中，我已經學會了如何成為一個有愛心和靈敏的人。而現在，我希望我可以給我的孩子們同樣的經歷，使他們成為一個更好的人。

說了這麼多，請考慮讓我們領養一隻您的小貓。我保證，我們會盡我們最大的努力給這個小傢伙一個充滿愛的家庭。我們願意接受任何品種的小貓甚至那些有狀況的。我們期待著聽到您的回覆。

您忠誠的，

Jack Thompson

例題中譯

1. Pur 救援是甚麼樣的組織？

A 它提供動物永久的家。

B 它提供動物不收費的醫療照顧。

C 它幫助狗和貓尋找新家。

D 它以合理的價格販賣動物。

2. 以下何者不是此文的主旨？

A 使人們瞭解動物收容所如何運作。

B 讓人知道此組織可以做到甚麼。

C 鼓勵人們從他們這裡領養。

D 讓人瞭解經營動物救援組織的困難處。

3. 寄養家庭是甚麼意思？

A 某人的永久家庭。

B 庇護所。

C 森林中的房子。

D 臨時的家。

4. Thompsons 為何想領養一隻貓？

A 因為他們沒有時間照顧狗。

B 因為小貓比狗親切。

C 因為 Thompson 先生想要他的小孩經歷他所經歷過的。

D 因為比較於狗，他們可以花費較少的關注力在貓上。

5. 接下來可能會發生甚麼事？

A Thompson 家庭將會被 Pur 救援訪談。

B Thompson 家庭可以選擇他們喜歡的貓。

C Thompson 家庭將改變想法並領養一隻狗。

D Thompson 家庭將會成為此組織的義工。

解答

1. C　2. D　3. D　4. C　5. A

解題

1. 在第一篇組織介紹的開頭已說明 Pur 救援是一個全志願非營利性的動物救援組織。此集團已經救出了數百隻寵物，並把他們安置在充滿愛的家園裡，因此解答為C。

2. 在兩篇文章中皆未提到經營動物救援組織的困難處，因此答案為D。

3. 寄養家庭代表著中繼之家，因此解答為D。

4. 在第二封信件的第三段中，Thompson 先生說明了他的成長經歷，並表明希望自己的孩子也可以有一樣的經驗，因此答案為C。

5. 在第二封回信中，Thompsons 表明了願意收養任何狀態的貓，因此 Pur 救援應會與 Thompsons 進行訪談。解答為A。

字彙

1. **Comprise** (v) 包括，包含，由…組成

   This play comprises 3 acts.

   這一齣劇是由 3 幕組成。

2. **Euthanize** (v) 安樂死

   Pete adopted his dog right before it got euthanized.

   Pete 在他的狗被安樂死前收養了牠。

3. **Alliance** (n) 同盟，聯盟

   There is a disagreement within the alliance about how to deal with this problem.

   對於如何解決這個問題，聯盟中出現了分歧。

4. **Enhance** (v) 提高，增加，增強

   You can enhance the flavor of the dish by using fresh herbs.

   你可以利用新鮮的香草來提升這盤菜的味道。

5. **Surround** (v) 環繞，圍繞

   When an idol visited Taiwan, the airport is always surrounded by exciting fans.

   當有偶像訪問台灣時，機場總是被興奮的粉絲包圍。

SAMPLE 14 Request for job interview

A Request for job interview

To: gianamills@exhibitdesign.com
From: cameronc@amail.com

Dear Giana,

I am responding to the advertisement in Mosters.com dated July 12th, 2015 for the role of Senior Sale Representative. You can find my resume at the following page. I have currently worked in my present position for 5 years and I feel like it is time for me to take on a bigger responsibility and challenge myself in an international **environment**. I feel I have all the necessary qualities that are needed to make the jump from regular sales to Senior Sales Representative in a firm such as yours. I realize that there will be much competition for a role in a company as large as exhibit design, but I feel that I would be the correct choice.

Thank you for your time today.

Sincerely,

Cameron Colin

Cameron Colin

email: cameronc@amail.com

HIGHLIGHTS OF QUALIFICATIONS

Excellent communication, time management and presentation skills

Effective, responsible and committed team player

Detail and deadline **oriented**, very organized

RELAVANT EXPERIENCE

CA Design Studio, Santa Clara, CA June 2010 – Present

Sales

Key activities:

- Present the company's activities and products to customers
- Develop an understanding of the customers' needs and offer the solution
- Prepare price quotation
- Analyze and prepare strategy for introducing new products to the market

Studies in American Language, San Jose, CA

August 2007 – May 2010

Student Assistant

Key activities:

- Provided coordination, promotion, and publicity for school events
- Facilitated translation services to students with English difficulty
- Managed office supplies

Hocus Pocus English School, Tokyo, Japan

April 2004– March 2005

English Instructor

Key activities:

- Helped students improve English writing and conversational skills
- **Consoled** students with difficulty in English
- Organized students' off-class special activities

EDUCATION San Jose State University, San Jose, CA

Bachelor of Science Degree in Business Administration with an **emphasis** in Marketing

August 2007 – May 2010

SKILLS Computer Skills: Proficient in MS Word, Excel, PowerPoint, Outlook, Adobe Acrobat

Languages: Fluent English and Japanese

B Response

To: cameronc@amail.com
From: gianamills@exhibitdesign.com

Dear Cameron:

Thank you for taking an interest in our company. After reviewing your resume, we think you might be suitable for this job. Though we have concerns that you do not hold a design-related degree, we think your previous experience with CA Design Studio might **make up** for the lack of education. And the billingual capability definitely will be a plus for the international sales role.

We would like to meet with you in person. Please contact our HR Department and set up an interview. We look forward to hearing back from you.

Best regards,

Giana Mills

Answer the Following Questions

1. What kind of a job is Cameron applying for?

A Junior sales

B Designer

C Senior coordinator

D Senior sales representative

2. Why does Cameron want a new job?

A The current job is too easy for her now.

B She thinks she is too busy.

C She wants to be in an international environment.

D She wants to earn more money.

3. What quality of Cameron attracted Exhibitdesign the most?

A She lived in Japan for a year.

B She is detail and deadline oriented.

C She holds a degree in business.

D She speaks 2 languages fluently.

4. According to the second email, what should Cameron do next?

A Quit her current job.

B Call the HR at Exhibitdesign to arrange an interview.

C Celebrate because she already got the job.

D Send Giana a thank you note.

5. According to the resume, which of the followings is NOT true?

A Cameron lived in Japan during her college years.

B Cameron speaks 2 different languages.

C From 2007 to 2010, Cameron works part time while finishing her degree.

D Cameron only has one formal working experience.

Answer

1. D 2. C 3. D 4. B 5. A

例文十四 自我推薦

A 要求面試

此至：gianamills@exhibitdesign.com
來自：cameronc@amail.com

親愛的 Giana，

我正在回應您於 2015 年 7 月 12 號登在 Mosters.com 上應徵高級銷售代表的廣告。您可以在接下來的頁面裡找到我的履歷。我已經在我目前的職位任職 5 年，我認為是時候讓我承擔更大的責任，並在國際化的環境裡挑戰自己。我覺得我具有在您的產業裡從一個一般銷售轉變為高級銷售代表的素質。我能體會在像 Exhibitdesign

這樣的大公司必定會有很多的競爭對手，但我相信我將會是您正確的選擇。

謝謝您今天的時間。

真誠的，

Cameron Colin

Cameron Colin

電郵 : cameronc@amail.com

資格亮點

- 優秀的溝通、時間管理和演講技巧。
- 有效的、負責任的和堅定的團隊合作精神
- 注重細節和期限，非常有條理

相關經驗

CA 設計工作室 , 聖塔克拉拉 , 加州　　2010 年六月 – 現在

業務

重點活動：

- 向客戶發表公司的服務及產品。
- 開發並瞭解客戶的需求並提供解決方案
- 準備報價單
- 分析並準備介紹新產品上市場的戰略

美語學習部門，聖荷西，加州　2007 年八月 – 2010 年五月

學生助理

重點活動：

- 協調、宣傳學校活動
- 提供有英語困難的學生翻譯的服務
- 管理辦公室文具

Hocus Pocus 英語學校，東京，日本　2004 年四月 – 2005 年三月

英語教師

重點活動

- 幫助增進學生的英語寫作及會話能力
- 安撫英語有困難的學生
- 整合學生的課後活動

教育　聖荷西州立大學，聖荷西，加州

工商管理學士學位，專供市場營銷

2007 年八月 – 2010 年五月

技能　電腦技能：

精通：MS Word, Excel, PowerPoint, Outlook, Adobe Acrobat

語言：流利的英文及日文

B 回覆

此至：cameronc@amail.com
來自：gianamills@exhibitdesign.com

親愛的 Cameron：

感謝您對本公司抱持興趣。回顧你的簡歷後，我們認為您可能很適合這份工作。雖然我們對於您未有設計相關的學位有顧慮，但我們認為以您之前在ＣＡ設計工作室的經驗應該可以彌補缺乏的教育部份。以及您的雙語能力將會是成為國際銷售員的加分。

我們希望與您見面。請聯繫我們的人力資源管理部門，並安排面試。我們期待著聽到您的回覆。

最好的問候，
Giana Mills

例題中譯

1. Cameron 是在應徵什麼樣的工作？

A 初級銷售

B 設計師

C 資深協調員

D 高級銷售代表

2. Cameron 為什麼想找新工作？

A 目前的工作對她來說太容易。

B 她認為太忙碌。

C 她希望能在國際化的環境。

D 她想要賺更多的錢。

3. Cameron 的什麼資質最吸引 Exhibitdesign?

A 她曾住在日本一年。

B 她注重細節及期限。

C 她擁有商學位。

D 她能說流利的兩國語言。

4. 根據第二封電郵，Cameron 接下來將做什麼？

A 辭掉現在的工作。

B 打電話給 Exhibitdesign 的人事並安排面試。

C 慶祝，因為得到了新工作。

D 寄感謝函給 Giana。

5. 根據履歷，以下何者為非？

A Cameron 在大學時住在日本。

B Cameron 能說兩種不同語言。

C 在 2007 到 2010，Cameron 在完成學業的同時也在打工。

D Cameron 只有一個正式的工作經驗。

解答

1. D　2. C　3. D　4. B　5. A

解題

1. 在第一篇電郵的開頭 Cameron 已表明是對高級銷售代表一直的興趣，因此解答為D。
2. 在第一篇電郵中 Cameron 表示希望能在國際化的環境裡挑戰自己，因此答案為C。
3. 在第二篇 Giana 的回信中提到 Cameron 的雙語能力將會是成為國際銷售員的加分，因此解答為D。
4. 第二封回信中，Giana 要求 Cameron 跟人事部安排面試，因此答案為B。
5. 由履歷我們可以瞭解 Cameron 大學是在美國就讀，因此解答為A。

字彙

1. **Environment** (n) 環境，周圍狀況

   Pollution is bad for the environment.

   污染對環境是不好的。

2. **Orient** (v) 等同 Orientate，調整，適應

   The class is oriented to non-English speakes.

   這個課程是為了非說英語者所調整的。

3. **Console** (v) 安慰，慰藉

   We tried to console her when her dog died.

   她的狗死後，我們盡力的安慰她。

4. **Emphasis** (n) 強調，重點，重要性

   This English course puts a great emphasis on writing skills.

   這個英文課程非常強調寫作能力。

5. **Make up** 補足

   In order to make up the lack of Spanish language ability, he takes online courses everyday.

   為了補足西班牙語能力的不足，他每天上網上課程。

SAMPLE 15 Debate

A Homework or not? NO!

Ever since kindergarten, or even preschool, learning seems non-stop to most of us. Homework, like a **status quo**, never leaves our sight. However, is homework really necessary? Shouldn't kids be enjoying their private time after school? Recent research says it all.

Many researchers have been doing studies and trying to find evidence to support the benefit of having homework, but are coming up **empty-handed**. Back in time while we were students, there was a period of time that som of us were struggling to concentrate on doing our homework while others are out there playing at the front yard. Also, the tension level between myself and my parents are skyrocketedly high. I for sure say that it **diminished** my interest in learning. I would say that the relationship between me and my family would be much better if school passed through the policy of not having homework.

I am glad to see that some educators have the exact same thought as myself and several schools adopt a no-homework policy across all grades in US. They try to make sure that students enjoy learning without stress. When kids have more time for social network and creative activities, their performance is similar to or even better than others, according to statistics.

B YES!

Homework or not? My vote goes to "Yes, homework!" Even though many studies have shown that many people are fearing that homework does more harm than good and causes **copious** amount of stress to students and parents, sometimes even teachers, we still believe that practice makes perfect. And by giving students homework, it helps them think independently.

Homework is a good way to shorten the distance between teacher and students. By discussing the assignments, students can easily bring up their problems outside of the text books to their teachers. Teachers can also use this time to understand their students deeper.

Although many people advocate that homework increases the distance between family members, my opinion is the opposite. If parents or family members are willing to spend time to doing homework with their children or siblings, it not only increases the effectiveness, but also shortens the distance between family members.

Also, homework is a great way to gain responsibility. Kids needs to get a sense of **punctuality**. And by forcing them to turn in homework on time, it helps to develop this matter.

For all the above reasons and many more that are not listed, I shall say that homework will only do good to children and should not be eliminated.

Answer the Following Questions

1. What does "STATUS QUO" mean ?

A An old saying

B The existing state of affairs

C A rumor

D The old adage.

2. According to the 1st article, why do people think kids should NOT have homework?

A Homework makes kids less responsible.

B Homework increases kids concentration.

C Homework increases the distanse between kids and their parents.

D Homework gives kids more social time.

3. According to the 2nd article, why SHOULD kids have homework?

A Kids can discuss their homework with friends and improve their social skill.

B Kids will be more responsible.

C Kids will depend on their parents more.

D Homework helps teachers know their students better.

4. Why is doing homework a great way to develop responsibilities?

A It helps develop a sense of.

B Kids will learn more from their surroundings.

C Kids will study with their friends and build up small society.

D Teachers give the agenda of when to do what.

5. According to the 2 articles, which of the following is NOT true?

A Homework makes kids more responsible.

B Homework does not shorten the distance between kids and their families.

C Teachers can get closer with their students if students bring their questions about homework to them.

D Most kids hate homework.

Answer

1. B 2. C 3. B 4. A 5. D

例文十五 辯論

A 應不應該有功課？不應該！

自從上幼稚園，甚至幼幼班，對我們大多數的人而言，學習似乎永無止盡。功課就像一個現狀，從未離開過我們的視線。然而，功課真的是必要的嗎？孩子們不是應該在放學後享受自己的私人時間嗎？最近的研究說明了一切。

許多研究人員一直在做研究，試圖找到證據來支持有功課的好處，但還是兩手空空。回想到過去我們還是學生時，我們有些人都經歷過有段時間，當大家都在前院玩耍時，自己卻在掙扎於專心做功課。另外，我和我的父母之間的緊張程度更是異常的高。我可以肯定的說，它減低了我學習的興趣。我要說的是，我與我家人的關係可能會好得多如果學校通過取消功課的政策。

我很高興地看到一些教育工作者和我有著相同的觀點，而且在美國的幾個學校所有年級實施了無家庭作業的政策。他們試圖確保學生喜歡學習，沒有壓力。當孩子們有更多的時間用於社交網絡和創作活動，根據統計，他們的表現與他人相同或甚至更好。

B 應該

家庭作業與否？我的票投給“是的，功課！”儘管許多研究顯示許多人擔心家庭作業弊大於利，且對學生、家長，有時甚至老師造成極大的壓力，但我們仍然認為熟能生巧。同時，通過給學生的作業，可以幫助他們獨立思考。

家庭作業是縮短師生間距離的好方法。通過討論作業，學生可以很容易地把他們教科書以外的問題帶給老師。老師也可以利用這個時間，更加的瞭解自己的學生。

雖然很多人主張功課增加家庭成員之間的距離，但我的想法是相反的。如果父母或家庭成員願意花時間與他們的子女或兄弟姐妹一同做家庭作業，這不僅增加了效益，而且也會縮短家庭成員之間的距離。

此外，功課是一個獲得責任感很好的方式。孩子們需要了解準時的感覺。迫使他們準時交出功課有助此事的進展。

基於上述及許多尚未列出的原因，我認為功課對孩子是很好的，不應該被淘汰。

例題中譯

1. “STATUS QUO” 是甚麼意思？

A 古人說

B 一個現狀

C 一個傳聞

D 一個古老的諺語

2. 根據地一篇文章，為什麼人們認為不應該有功課？

A 功課減少孩子的責任感。

B 功課增加孩子的注意力。

C 功課增加孩子與家長的距離。

D 功課讓孩子有更多的社交時間。

3. 根據第二篇文章，為什麼小孩應該有功課？

A 小孩可以與朋友討論功課以增加他們的社交能力。

B 小孩將更有責任感。

C 小孩將更依靠父母。

D 功課幫助老師更瞭解他們的學生。

4. 為什麼寫功課是一個開發責任心的好方式？

A 它幫助開發規律性。

B 孩子可以從周圍學習更多。

C 孩子可以與朋友一起學習並建立小社會。

D 老師會給課表說明什麼時候做什麼事。

5. 根據這兩篇文章，以下何者為非？

A 功課使小孩有責任心。

B 功課不能縮短小孩與家人的距離。

C 若學生把功課上的問題帶給老師將拉近老師與學生。

D 大部份的小孩討厭功課。

解答

1. B　2. C　3. B　4. A　5. D

解題

1. 依照字面翻譯，解答為B。
2. 第一篇文章第三段中提到功課增加了孩子與父母的距離，因此答案為C。
3. 在第二篇中所提到功課的好處，在最後一點裡提到功課可以增加責任心，因此解答為B。
4. 第二封第四段中提到功課可以讓孩子在固定的時間做固定的事，這便是規律性，，因此答案為A。
5. 在兩篇文章中都未提及到學生的想法，因此解答為D。

A 字彙

1. **Status quo** (n) 現狀

   People often tend to want to maintain the status quo to eliminate problems.

   人們通常傾向維持現狀以減少問題。

2. **Empty handed** 空手，無結果的

   It always brings down the scientists when the result comes up empty handed.

   當結論是無結果的，科學家通常心情低落。

3. **Diminish** (v) 降低，減少

   The drug's side effects should be diminished over time.

   這個藥物的副作用將會隨時間降低。

4. **Copious** (adj) 豐富的

   The storm produced a copious amount of rain.

   這個暴風雨帶來了豐沛的雨。

5. **Punctuality** (n) 準時

   Kids should be trained with punctuality which will be a great help after they grow up.

   應該教導小孩準時，這將成為他們長大後很好的助力。

英語學習—生活·文法·考用—

定價：NT$369元/K$115元
規格：320頁/17＊23cm/MP3

定價：NT$380元/HK$119元
規格：320頁/17＊23cm/MP3

定價：NT$349元/HK$109元
規格：352頁/17＊23cm

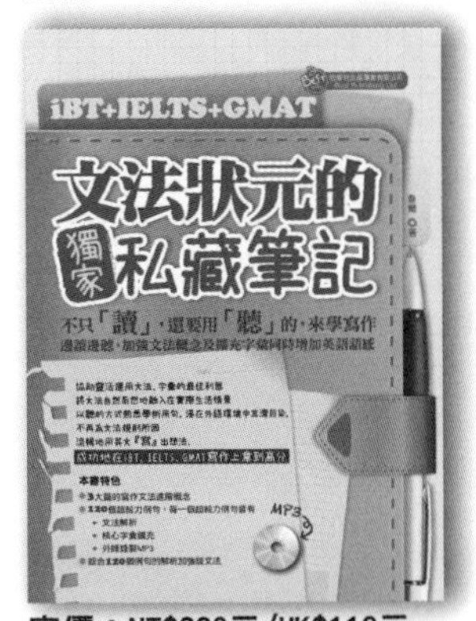

定價：NT$380元/HK$119元
規格：288頁/17＊23cm/MP3

定價：NT$329元/HK$103元
規格：352頁/17＊23cm

定價：NT$349元/HK$109元
規格：304頁/17＊23cm

定價：NT$380元/HK$119元
規格：352頁/17＊23cm

定價：NT$369元/HK$115元
規格：304頁/17＊23cm/MP3

定價：NT$380元/HK$119元
規格：304頁/17＊23cm/MP3

Learn Smart！050
國貿與新多益 一魚兩吃
「考用」與「職場」大結合迅速掌握「商用英文書信」要訣!
作　　者　洪婉婷
封面構成　高鍾琪
內頁構成　華漢電腦排版有限公司

發 行 人　周瑞德
企劃編輯　陳韋佑
校　　對　陳欣慧、饒美君、魏于婷
印　　製　大亞彩色印刷製版股份有限公司
初　　版　2015 年 9 月
定　　價　新台幣 349 元
出　　版　倍斯特出版事業有限公司
電　　話　(02) 2351-2007
傳　　真　(02) 2351-0887
地　　址　100 台北市中正區福州街 1 號 10 樓之 2
E-mail　best.books.service@gmail.com

港澳地區總經銷　泛華發行代理有限公司
地　　址　香港新界將軍澳工業邨駿昌街 7 號 2 樓
電　　話　(852) 2798-2323
傳　　真　(852) 2796-5471

國家圖書館出版品預行編目(CIP)資料

國貿與新多益 一魚兩吃「考用」與「職場」大結合迅速掌握「商用英文書信」要訣! / 洪婉婷著. -- 初版. -- 臺北市 : 倍斯特, 2015.09 面 ; 公分. -- (Learn smart! ; 50)
ISBN 978-986-91915-3-1 (平裝)
1.商業書信 2.商業英文 3.多益測驗
493.6　　104015655

Simply Learning, Simply Best

Simply Learning, Simply Best